Management of Solar Energy Technologies and Women Empowerment: A Case of Women Barefoot Solar Engineers of India

Management of Solar Energy Technologies and Women Empowerment: A Case of Women Barefoot Solar Engineers of India

Dr G. Valentina

GenNext Publication

ISBN : 978-93-8022-278-3

Price : ₹ 450

First Published, 2016

Published by

Gennext Publications
5, Ansari Road
Daryaganj, New Delhi-110002
Phones: 9811692060
E-mail: gennext@hotmail.com

Printed at: G. Print Process, Delhi.

Cataloging in Publication Data—DK
Courtesy: D.K. Agencies (P) Ltd. <docinfo@dkagencies.com>

Valentina, G., author.
Management of solar energy technologies and women empowerment / Dr G. Valentina.
pages cm
ISBN 9789380222783

1. Rural electrification—India—Thamingula. 2. Rural electrification—India—Poosalapalem. 3. Women in energy development—India—Thamingula. 4. Women in energy development—India—Poosalapalem. 5. Solar power plants—India—Thamingula—Management—Citizen participation. 6. Solar power plants—India—Poosalapalem—Management—Citizen participation. 7. Poor women—Employment—India—Thamingula. 8. Poor women—Employment—India—Poosalapalem. I. Title.

DDC 333.7932095485 23

Contents

एक कदम स्वच्छता की ओर

सत्यमेव जयते

ग्रामीण विकास मंत्रालय
ग्रामीण विकास विभाग
भारत सरकार
कृषि भवन, नई दिल्ली-110114
Ministry of Rural Development
Deptt. of Rural Development
Government of India
Krishi Bhavan, New Delhi-110114

Foreword

Access to energy is iniquitous in India. The ability of the poor to access grid connected power is limited by constraints, imposed by geography, wealth and power. The solar photovoltaic power (SPV) presents an option to ensure that at least basic electric power need of poor households are met. However, in the past, high cost of photovoltaic cells, inefficiency of energy conversion and difficulties of storing power to be used when sunlight are not available, made this option difficult to choose. Recent improvements in technology have increased efficiency of photovoltaic cells while reducing their cost. Alongside large projects for generating solar photovoltaic power which will help to reduce carbon footprint in energy generation and distribution, new technology also offers the possibility of decentralised production of power and its, use. It can also be leveraged to generate decentralised production opportunities which can, assist generation of employment in villages.

2. This book is a description of the work .done, by NIRD in empowering rural women and building their capacities to produce lighting solutions using appropriate technology. Some of the implications of this project have been 'examined here. The extent to which rural women can be trained in new technology, their ability to market the product among rural population as well as mobilising local communities for absorption of technology are some of them. This offers a proto-type for dissemination appropriate technology, generating a demand for new products and empowering family entrepreneurs

to develop self-sustaining business. This book can be a starting point for further exploration of issues involved in NIRD's efforts to work with rural women to create female entrepreneurs in appropriate technology.

(Rajeev Sadanandan)

Acknowledgements

The book Management of Solar Energy Technologies and Women Empowerment: A Case of Women Barefoot Solar Engineers of India is a Process Study (Monograph). I am privileged to work with and enjoy the support of a vast number of officials in initiating the project and also successfully bringing it to its logical end by meeting its objectives. The task of compiling this book has been challenging in the event of addition of material on the experiences absorbed by the village. I accord my sincere thanks to all the officials and the village community who cooperated with me in commissioning this project.

It would not have been possible for me to complete this book without the constant encouragement and support accorded to me by my husband Dominic a Pilot with Emirates, my parents, Mr.Franklin Bullabbai and Mrs.Vijayam and the untiring efforts and prayers of my colleagues and our institution. I gratefully acknowledge their support in all my endeavors till this day.

Dr. G. Valentina

Executive Epitome
Management of Solar Energy Technologies and Women Empowerment: A Case of Women Barefoot Solar Engineers of India

Dr. G. Valentina

India is a very large country with nearly 70% of its population living in the rural areas, which almost means 6 lakh villages in the country. Of these villages nearly 25,000 villages are unelectrified and about 18,000 villages cannot be electrified at all due to the geographical location of the villages which are dispersed and positioned on the hilltops. And all such villages are inhabited by the tribals and this population is devoid of an access to household electrification despite of six decades of independence of the country. Contextually, the government of India has made umpteen attempts to electrify these villages by connecting the villages to the grid power but the systems were not sustainable due to lack of operation and maintenance services of the systems. Secondly because of geographically long, and far flung distances and spatial disaggregation of the habitat, whenever there is a system failure they are unattended to due to lack of connectivity and access into the village. The end result is that the tribal population depend on Kerosene to illuminate their

Households and the expensive infrastructure installed in the villages stare out as skeletons depicting huge public financial investments.

In the last decades ministry of new and renewable energy has provided solar lanterns, home lighting systems and street lighting systems to the tribals at subsidized rates and almost free of cost, but the tribals do not know to use the lanterns and maintain the home lighting and street lighting systems in times of their breakdown, as a technical expert with knowledge to service the systems (locally) is not available in the village. In order to overcome this problem empowering the local rural communities on the technology is found to be the only option to sustain the technology.

Accordingly, four illiterate women living in the vicinity of NIRD Rajendranagar slums were selected and trained in solar energy technologies. The women were illiterate and could not speak any other language except for their local dialect. Despite their illiteracy they learnt to produce the systems where in technology was the only language through which the capacity building was done. During the training the women were taught to produce solar lanterns, solar home lighting systems, solar street lights and installation of solar power plants. These four ladies after their return from training have installed 5 Kw power plant, a Solar House at RTP NIRD without seeking the guidance or assistance of an expert. They have also fabricated more than 500 solar lanterns and marketed them to the NGOs and VIPs of the country, which have no reference for maintenance problems till date. In addition to the lanterns they have also assembled and fixed stand alone streetlights on a project basis in the states of Andhra Pradesh and Orissa, and also take care of the maintenance of the installed systems. The rural energy workshop installed by the BSEs along with the 1kw power plant is used to fabricate the solar systems and train the individuals on entrepreneurship in solar energy technology. Today they are established and registered themselves into Women Barefoot Solar Engineers Association (WBSEA).

The association comprises of four women namely Chennamma, Yellamma, Kalawathi and Zahida. Chenamma is 45 years old and worked as a stonecutter earning less than Rs 1000/month. She is married to an unskilled daily wage laborer and has 4 children. She only spoke Telugu with little or no knowledge of Hindi.

Zaheda is 35 years old and worked as a housemaid washing clothes and dishes. She was earning Rs 500/ month and raising 4 children, she is 4th standard pass, divorced and living with her mother. She speaks a bit of Hindi and fluent Telugu. Yellamma also worked as a stonecutter and is completely illiterate. She has 3 children and her husband too works as a stonecutter. Kalavathi is also illiterate, married and has 3 daughters. She was working as a housemaid and was earning a paltry amount of Rs. 500 per month.

Being encouraged by the interventions of WBSEA, the Commissioner Tribal Welfare, Government of Andhra Pradesh, has extended a project to solar electrify 128 households at Paderu mandal in Thamingula and Pusalepalem villages. After scrutinizing a detailed budget proposal and taking the cost of home lighting into consideration, the Commissioner for Tribal Welfare has released an amount of Rs. 23, 34,371 lakhs in Phase-1, towards training for four Barefoot Solar engineers, assembly and installation of home lighting systems for 128 households, establishment of one Rural Electronic workshop and Village Energy and Environment Committee training. Each home lighting system costed Rs.13, 500 which provided a 37-Watt Solar Panel, 2 CFL tube lights of 9 Watts each, a DC fan and a black and white TV connection. In PHASE-II a budget for Rs.5, 56,251 lakhs, was drafted towards the installation of a 1 KW power plant at Thamingula village, which will provide 10-12 streetlights and a small power pack to install a community Television.

As soon as WBSEA bagged the project, the solar engineers from RTP, NIRD, visited Thamingula and Pusalepalem villages of Paderu ITDA division, Visakhapatnam district, Andhra Pradesh. They selected four young tribal girls, through a participatory approach with the sole idea of providing them training on solar technologies.

The previous experiences of investments in developmental activities have established unsustainability of the programmes (due to lack of servicing facilities upon its breakdown), and in order to make the programme successful and sustainable, the selected Tribal girls have been sent for training from 22nd of August 2005 and are taught to fabricate solar lanterns, assemble home lighting systems and street lights so that all the breakdowns and failures of the systems in the future can be addressed to by the trained Tribal Girls themselves, who after the completion of the training are called as Barefoot Solar Engineers (BSEs). The names of the four trainee BSEs are:

1. S.Papayamma from Thamingula village,
2. S.Pravalika from Thamingula village
3. S.Kumari from Pusalepalem village and
4. S.Santhoshi from Pusalepalem village.

In addition to the selection of the trainees for capacity building on Solar Technologies, Village Energy and Environment Committees (VEEC), comprising of the Self Help Group Leaders and the Panchayat Members, have been formed which will monitor the installation and maintenance of the systems in the future.

A unique feature of this project is that the villagers have agreed to contribute an amount of Rs.1000/ household at the time of installation of the home lighting system and Rs.100 every month/house hold (after the installation of the home lighting systems) as user charges. The WBSEA team has opened an account in Union Bank of India at Chintapalli, in the name of VEEC, and based on the nomination of the villagers two individuals from each village have been selected and given the signatory powers to operate the accounts with a rider that only with the approval of WBSEA president the withdrawals can be carried out. The amount is collected with a prudent thought that it will be utilized for purchasing new batteries after 5 years as the batteries weaken and payment of salaries for the four trained Tribal Barefoot Solar Engineers.

In six months the four trainees completed the training and have also fabricated 132 home lighting systems along with four solar engineers of WBSEA, RTP, NIRD. After the completion of training, the trainees moved to their villages and installed 132 home lighting systems in Thamingula and Pusalepalem villages and the villagers have started using the solar systems.

Reaping satisfactory results the government of Orissa extended another project to solar electrify 30 households in Tingnaput village, Koraput district. Here also the 4 barefoot solar engineers have replicated the model enunciated at Paderu in Vishakhapatnam district. In this project also four illiterate women belonging to the khond tribes have been selected from the village and sent for training. They are Nishika, Raso, Talso and Bundei. They are the members of the self help groups. These Tribal women have no basic education, yet upon training for 6 months the women learnt the assembly of technology and also carry out maintenance of the systems in the village. The BSEs get a salary of Rs.12-15000/ from the user charges that are paid by the households.

In addition to the repair and maintenance of the systems the trained barefoot solar engineers from paderu are involved in manufacturing solar lanterns. Being encouraged by the solar electrification project the villages in the vicinity of the project villages have started making a demand for the lanterns and this has opened an avenue for the trained barefoot solar engineers to start manufacturing the solar lanterns in the renewable energy workshop in the village itself. Apart from producing and selling the lanterns the BSEs have also started renting the lanterns to the other villagers so as to utilize the lanterns in the night times and they charged a rent of Rs.30 / day. On an average they are able to earn Rs.120-150 per week.

By now the NIRD BSEs have installed projects on a turnkey basis in the following order.

- The technology institutes like the Engineering colleges have taken orientation training from the BSEs for one week. This has generated an additional income of Rs.5, 000 for 3 participants for three days.
- Being encouraged an NGO by name Surendra Solar Society in Haryana has invited WBSEA to build capacity of the women from their villages, as the villages in Haryana are facing an acute power failure and feel the need to adopt alternate energy systems.
- Home lighting in Paderu, Tingnaput village- Orissa
- Installation of 5 kw power plant in SAHARA NGO -Lucknow
- Capacity Building - Surindra Society, Haryana

- Electrification of 200 Tribal villages by Eastern Power Company
- Training for Nuzuvedu Engineering College students
- Training for Shadan College -Engineers Without Borders
- Conducted Training for 1000 participants; Including TOT to Rural Youth, AIEEE, Anwar -Uloom Engineering Students
- Solar Street Lighting in NIRD Action Research villages
- Marketing of lanterns, Home lighting systems, study lamps, LED based lamps -in partnership with PRAKRUTHI NGO
- Dissemination of knowledge and information in National level & International level seminars Including -JAPAN RENEWABLE ENERGY INSTITUTE - TOKYO

Today they are able to manage the solar energy technologies all by themselves and the women barefoot solar engineers' stand empowered. The present book is an attempt to give a detailed process study (Travel Logue) of the system developed by NIRD.

Introduction

Energy plays a central role in improving the quality of lives, health, education, communication, access to information and also empowerment of women. Energy is a key requirement for addressing challenges on poverty and climate change. As per the existing scenario lack of access to modern energy has placed 44% rural households to use kerosene to light their houses (364 million people) however in urban India -spread of electricity is 93%.

Statistics reveal that there are 80,000 unelectrified villages in India out of which 18,000 are remote villages. And those villages that have been provided grid power receive less than 6 hours supply which means that there is a lack of clean and safe lighting.

Therefore, lot of limitations are foreseen in the expansion of electricity grid which has stood as a formidable challenge to planners & development practitioners.

Pondering on the lacuna, when we probe (question) what's the way out? The answer indicates a need for enormous efforts to ensure energy access for all. Most parts of India have 300 - 330 sunny days in a year which is equivalent to over 5000 trillion kwt/h per year. This is more than India's total energy consumption per year thus solar energy can be considered as an alternative challenge to address the issue.

Against the existing scenario adopting the solar systems will facilitate many advantages like a substantial reduction of local pollution, reduction in the GHG (greenhouse gas) emissions, reduction in the cost (after initial investment of the solar systems), a good source of employment, ideal for remote locations, low transmission costs etc. However, disadvantages are limited like there is of a vast land tracts of land usage while providing solar grid power and limited availability of the power.

A technology demonstration unit was parked in Rural Technology Park at the National Institute of Rural Development with the modus operandi to train illiterate women in management and sustainability of the technology, transfer of technology to rural areas and empower communities in sustaining technologies through Village Energy and Environment Committee.

Pilot study findings revealed that the tribals use Kerosene for electrifying their housing units and procure kerosene from the PDS in the village. The tribals belonged to Bhagatha sub tribe and their livelihood dependency is on forest collection of adda leaves, fruits & by products.

The per capita income is not more than Rs. 1000/- per family per month while previous attempts of electrification in the village were unsustainable.

Initiation of solar electrification project in the project area was done in three steps firstly, people's participation for sustainable solar electrification programme, demonstration of technology, selection of trainees for capacity building programme, secondly initiation of sustainability measures, formation of Village Energy & Environment Committees & thirdly community contribution for project, an installation amount of Rs.1000 per household and user charges of Rs.100 per month for utilizing electricity.

Against all appreciation of the installation of the system, there are certain problems that have been encountered in the study area encouraging conflict of thoughts due to the flow of negative information in the community on the functioning of the systems which was successfully addressed by conducting the awareness programmes.

In the process, the installation and management of the technology in the village has led to the women empowerment where the women themselves organized the transportation of 20 ton material to project village, booked vehicles to carry the material, packed the material taking care to avoid all damages and assembled and Installed 132 solar systems in the village in 4 days along with installation of a half a kilo Watt solar power plant. In addition to the technical part of the empowerment the women were empowered in terms of employment, social development, livelihood development, awareness, decision making capacity, leadership, financial empowerment, problem solving abilities etc.

In the study it was interesting to note peoples struggle for electrification, initiatives taken by women, SHGs intervention in getting the credit for solar systems, motivation provided by the children to the parents to electrify their houses etc.

Problems encountered in the project were very few but very challenging as in the collection of user charges and misuse of the systems.

At the end of the project period, the results were encouraging and the impact was legible in terms of a clear positive health effect on the households, children's ability to do their school homework and study in the evenings has increased and also claims of achieving better marks in school examinations were evident. TV has brought entertainment, news and educational programmes to villagers, they have also become more aware of their social standing, available electric light has been helping women to cook food in the evening, installation of fans makes heat more bearable and helps avoid insect and mosquitoes bites, women are able to work for extended hours in the evening, cleaning paddy, tamarind and the turmeric pod, it was encouraging to note that some villagers have expressed their desire for addition of more household appliances such as coolers and refrigerators through the solar systems.

I

Solar Electrification in Rural Tribal Areas

1. Introduction

Industrialization, commercialization and growing population have exerted much pressure on the dwindling energy resources in India. Today world over energy resources are becoming a scarce commodity and energy efficiency is assuming much importance. On the other hand around 23 per cent of 5, 86,000 villages in India and 56 percent of 138 million households in India do not have electricity connectivity. Difficult and inaccessible terrain makes it unviable to extend electricity distribution network to such areas. This inaccessibility is mainly due to remote locations of the habitations of population. Hence it has become a Herculean task for various governments to bring such scattered population into the ambit of electrification. There were several attempts and financial investments made by the governments in various plan periods to provide electrification to the rural population in the far flung hilly and remote areas especially for the tribal settlements but it was all a futile exercise. Lying of electrical infrastructure from the central grid to the vastly located geographical terrains was not feasible technically and economically. Overlooking these hurdles attempts have been made to electrify remote and inaccessible villages but due to the naturally occurring high velocity winds in the altitudes, the erected polls failed to withstand the pressure resulting in its uproot. Consequent to the failure of the systems, neither the government nor the local population could pull their gots to evaluate the implemented programmes or to revive the system. However loss is a loss either to the government or to the society, in addition to the disappointment and humiliation.

Therefore efficient and reliable energy supply stood out as a challenge and turned out to be critical in the rural belts. On the event of non-applicability of the non renewable energy systems for electrification, today the renewable energy sector provides maturity in its technology application, which has grown in volumes. Consequently our attempts in India in the last decades have shown that due to lack of aftercare- follow up -maintenance, these systems have become defunct in many places.

India being a large country with nearly 70% of its population living in the rural areas which almost means 6 lakh villages in the country of which nearly 25,000 villages are unelectrified and about 18,000 villages cannot be electrified at all due to the geographical location of the villages which are dispersed and positioned on the hilltop, such villages are occupied by the tribals and this population is devoid of any access to electrification despite of six decades of independence of the country. The government of India made umpteen attempts to electrify these villages by connecting the villages through the grid power but the systems were not sustainable due to lack of operation and maintenance.

Reports support that long geographical distances, lack of attention to the systems in times of failure, lack of connectivity and access into the village have been the main reasons.The end result is that people depend on Kerosene to illuminate their households despite the availability of grid power in the village.

Thus the expensive infrastructure installed in the villages stand as skeletons depicting huge investments of public money .In addition the ministry of new and renewable energy has provided solar lanterns, home lighting systems and street lighting systems to the tribals at subsidized rates and almost free of cost. but the tribals did not know how to use the lanterns and maintain the home lighting and street lighting systems upon breakdown and an expert having the knowledge to service the systems locally is also not available. In order to overcome this problem empowering the local rural communities on the technology is the only option to sustain the technology.

One of the viable alternatives here is solar energy. These technologies demonstrate the potential and opportunities of tapping solar energy as an alternative to non-renewable energy systems. Keeping this in view, NIRD has evolved a concept of building up capabilities among the local community to assemble, install, operate and maintain these solar energy systems as part of the projects for transfer of technology in Rural Technology Park by empowering four semiliterate women on the assembly of solar energy technology.

Fig. Thamingula Village

1.1 Empowerment of WBSEA Women

Four illiterate women near NIRD Rajendranagar slums were selected and trained in solar energy technologies. The women were illiterate and could not speak any other language except for their local dialect. Despite which they learnt the technology to produce the systems where in technology was the only language through which they were empowered in the assembly of the technology. During the training the women were taught to produce solar lanterns, solar home lighting systems, solar street lights and installation of solar power plants.

These four ladies after their return from training have installed 5 Kw power plant at RTP, Solar House NIRD, without seeking the guidance or assistance of an expert. They have also fabricated more than 500 solar lanterns and marketed them to the NGOs and VIPs of the country, which have no reference for maintenance till date. In addition to the lanterns they have also assembled and fixed stand alone streetlights on a project basis in the state of Andhra Pradesh and Orissa, and also take care of the maintenance of the systems. The rural energy workshop installed by the BSEs along with the 1kw power plant is used to fabricate the solar systems and train the individuals on entrepreneurship in solar energy technology, and they have registered themselves into Women Barefoot Solar Engineers Association (WBSEA).

Today as an association, they handle their own accounts and maintain their records. They are Chennamma; Yellamma, Kalawathi and Zahida. They are in the age group of 40-45 years, worked as unskilled daily wage laborers (stonecutters/ house maids), earning less than Rs 1000/month.

Being encouraged by the interventions of WBSEA, the commissioner for Tribal Welfare, government of Andhra Pradesh, has provided a project to solar electrify 128 households at Paderu mandal in Thamingula and Pusalepalem villages. After scrutinizing a detailed budget proposal and taking the cost of home lighting into consideration, the commissioner for Tribal Welfare has released an amount of Rs.23, 34,371 lakhs in Phase -1,towards training for four Barefoot Solar engineers, home lighting systems for 128 households, one Rural Electronic workshop and Village Energy and Environment Committee training.

Each home lighting system costed Rs.13, 500/- which provided a 37-Watt Solar Panel, 2 CFL tube lights of 9 Watts each, a DC fan connection and a black and white TV connection. In PHASE -II the project involved a budget for Rs.5, 56,251/- lakhs towards the installation of a 1 KW power plant at Thamingula village which will provide 10-12 streetlights and a small power pack with which a community Television can be installed.

In addition to the selection of the trainees for capacity building on Solar Technologies, Village Energy and Environment committees (VEEC), comprising of the Self help group leaders and the panchayat members, have been formed which will monitor in the installation and maintenance of the systems in the future. A unique feature of this project is that the villagers have agreed to contribute an amount of Rs.1000 / household at the time of installation of the home lighting system, and Rs.100 every month /house hold after the installation of the home lighting systems as user charges. The WBSEA team has opened an account in Union Bank of India at Chintapalli, in the name of VEEC and based on the nomination of the villagers two individuals from each village are given the signatory powers to operate the accounts and at the behest of WBSEA president only the withdrawals can be carried out. The amount so collected and deposited in the bank, from the villagers is utilized for purchasing new batteries after 5 years as the batteries weaken, and payment of salaries for the four trained Tribal Barefoot Solar Engineers.

The four trainees have completed six months of training and have also fabricated 128 home lighting systems along with four solar engineers of WBSEA, RTP, NIRD in the phase -I of the project. After the completion of the training the trainees have moved to their villages and installed 128 home lighting systems in Thamingula and Pusalepalem villages by 22nd of March 2006 and the villagers have started using the solar systems. Now with the completion of phase -I, WBSEA is ready to take up phase -II of the project.

Reaping satisfactory results, the government of Orissa extended another project to solar electrify 30 households in Tingnaput village, Koraput district. Here also the 4 barefoot solar engineers have replicated the model enunciated at Paderu in Vishakhapatnam district. In this project also four illiterate women belonging to the khond tribes from the village to be electrified are selected and put on training. They are Nishika, Raso, Talso and Bundei. They are the members of the self help groups. The Tribal women have no basic education yet upon training for 6 months the women are able to learn the technology and take care of the maintenance of the systems in the village. The BSEs get a salary of Rs.12-15000/, from the user charges that are paid by the households.

In addition to the repair and maintenance of the systems the trained barefoot solar engineers from paderu are involved in manufacturing solar lanterns Being encouraged by the solar electrification project the villages in the vicinity of the project villages have started making a demand for the lanterns and this has opened an avenue for the trained barefoot solar engineers to start manufacturing the solar lanterns in the renewable energy workshop in the village itself. Apart from producing and selling the lanterns the BSEs have also started renting the lanterns to other villagers that want to utilize the lanterns in the night times. They have started charging an amount of Rs.30 for renting out the lanterns. On an average they are able to earn Rs.120-150 per week.

The technology institutes like the Engineering colleges also have taken orientation training from the BSEs for one week's time. This has generated an additional income of Rs.5, 000 for 3 participants for three days.

Being encouraged an NGO by name Surendra Solar Society in Haryana has invited WBSEA to build capacity of the women from their villages, as the villages in Haryanma were facing an acute power failure and feel the need to adopt alternate energy systems. The four women called the women barefoot solar engineers today are empowered and the sustainability of the enterprise is simply dependent on them. Four women selected for training base on their attitude to work, take responsibility and learn and technically being.

During their training at SWRC -Tilonia, they were taught a single method of assembling a circuit therefore any new circuit confused them initially and they could not work on such systems. Today the market is flooded with umpteen number of systems which can work more efficiently than the existing one on which they are working but they are able to understand the changes and developed the skill to address them on par with the latest technology, there by establishing the fact that the WBSEA BSEs are interested in Research and Development also and do not go by conventional methods of working for the systems though the BSES are not educated.

The BSEs are now getting a standard salary package. This has raised their standard of living and they are able to get their children educated and get the best medical help for themselves and their families. They also take decisions in the family as regards the children's education, health, purchase of items for the family, children's marriage etc.; this shows an impact of being empowered after the skill development programme at NIRD.

There is no tab over the finances that they want to spend; they do not give an account of the expenditure to their husbands who also do not ask them for it. They apply their minds when a job is allotted to them and procure the items on their own as they have knowledge on the components to be used. Initially the BSEs had an attitude of passing the buck and never took any responsibility but today they are confident and volunteer to take the responsibility.

Though they very backward in marketing they are able to sell the products successfully. Now the BSEs are also educated and can understand all conversations and will be able to provide the costing details of the technology and sort out the maintenance problems of the systems by themselves.

They cannot understand the credit and debit aspects of the banking process but they do remember the amount withdrawn and deposited at each instance and they are quick to calculate and give an estimate of the amount expended and the balance that could be available in their account, today they are of the feeling that the society is theirs and they promise to sustain it until the end.

The president works as a socialization agent and constantly helps the other BSEs to check their behavior and adapt to the circumstances in demand. However, educated trainees with zeal to work are the need of the hour to bring in a mix to promote efficiency in the society. There is also a need for a technical assistant to lead the group and bring in fresh learnings with the research and development taking place in the technology across the world.

The project was initiated with the following objectives:

1.3 Objectives

- To promote application of solar energy for rural development by training and networking semi-literate poor rural women from non-electrified villages/hamlets as barefoot solar engineers (BSE), establishing a training center exclusively for this purpose.

- To establish community-managed rural energy systems and powerhouses that work without dependency on exogenous skills and technicians, demonstrate that the communities can do it, and exhibit solar electrification as a viable and cost-effective alternative.
- To showcase solar technology as a promising income generating venture for the rural poor by fabricating charge controllers and inverters and assembling solar lanterns and fixed power packs in a Rural Electronic Workshop; and marketing the finished products and related services.
- To apply and receive funds from relevant sources and undertake turnkey assignments and utilize the proceeds towards furtherance of the objectives of the society.
- To identify, contact and encourage existing organizations in the field and develop collaborative linkages with them.
- To undertake process documentation of the 'barefoot approach' of rural electrification and use the audiovisual outputs for wider replication of the idea.
- To collaborate with NIRD for promotion of renewable energy sources in general and solar energy in particular through activities including organization of seminars and workshops aimed at sensitizing policymakers and panchayat representatives about the effectiveness of the barefoot approach.

Fig. Pasalepalem Village

1.4 Transfer of Technology

The area chosen for the transfer of technology has been decided by the A.P.Tribal power company. The selection of the area is based on the rules of MNRE which promulgated the provision of solar electrification to a village which is 7 km away from the last electrified village. Based on this criterion the two villages Thamingula and Poosalapalem are selected for solar electrification in Paderu mandal.

1.5 Methodology

The methodology involved in the transfer of technologies is in two phases. Initially community participation and capacity building on the technology is done for the community. Following which the trainees are selected and provided a detailed training to handle the technology. The whole process takes nearly six months to empower the trainees on the technology .With the completion of the training the systems are installed in the villages which are serviced in the village by the BSEs itself.

2

Conception of Barefoot Engineers for Rural Solar Electrification Project

2. Introduction

With the bourgeoning idea of establishing a technology park at the Rural Technology Park, National Institute of Rural Development, solar energy technology was considered as one of the technologies to be established at the park. The modus operandi of the technology was to train illiterate women in management, transfer of technology and sustainability of the technology in the rural areas and empower the communities in sustaining the technologies through capacity building. These primary interventions are based on the experiences of the Barefoot College in Tilonia which has won the name in building capacity and electrification to the rural poor in literally 600 villages in and around Rajasthan and North Eastern Areas. This gave way to the formation of a committee consisting of Shri Bunker Roy, Director General NIRD Shri Lalith Mathur, Deputy Director General Shri Chinmay Basu, Secretary Rural Development and functionaries from MNES on May 23, 05 which culminated in evolving a strategy to develop entrepreneurship in Solar energy technology.

2.1 Selection of Trainees for Solar Electrification Programme

On May 28, 2005 the Mahila Mandali of NIRD presented a group of poor ladies before the committee comprising of the Director General, Mr. Bunker Roy and Mrs. Director General. These women were living in the slums near NIRD. Only women with a potential to work and learn the assembly of solar energy technologies were selected. The committee selected a group of four women working as casual laborers in NIRD campus. They are Chennamma, Yellamma, Kalavathi and Zaheda. Chennamma and Yellamma are sisters and were working as quarry workers.

Chenamma is an illiterate and 45 years old and was working as a stonecutter earning less than Rs. 1000/month. She is married to an unskilled daily wage laborer. She has 4 children. She only spoke Telegu with little or no knowledge of Hindi. Zaheda is 35 years old and worked as a housemaid washing clothes. She was earning Rs 500/month and raising 4 children. She is 4th standard pass and is Divorced and living with her mother. She speaks a bit of Hindi and fluent Telegu. Yellamma also worked as a stonecutter and is completely illiterate. She has 3 children and her husband too works as a stonecutter. Kalavathi is also illiterate, married and has 3 daughters. She was working as a housemaid and was earning a paltry amount of Rs.500 per month.

The women were explained the purpose of their selection and being sent to the barefoot college and the ensuing responsibilities that they have to shoulder.

The ladies agreed wholeheartedly to undergo six months training at Rajasthan - Tilonia and their families also accepted the proposal and allowed them to attend the training programme. The young professionals of NIRD Mr.Ashwini Kumar and Mr, Vinay took the lead and dropped the ladies for training at Tilonia in July 2004.

The new experiences and new places made the ladies little nervous initially but soon they coped up and started concentrating on the work. During that period in Tilonia they learnt how to live and work together with people from other cultures and other countries like Africa, Afghanistan and Sierra Leone, through sign language and practical work because their Hindi was poor.

They learnt how to fabricate charge controller, invertors, solar lanterns and lights, install fixed solar units in individual houses and establish a Rural Electronic Workshop. Subsequently the women were totally involved in the work. In one occasion the father of the trainee (s) (Yellamma & Chennamma) expired but they preferred to continue with the training until its completion. Such was their commitment and sincerity for training and entrepreneurship.

By December 2004, five months training was completed and the women returned to Hyderabad being christened as women barefoot solar engineers. They came back to NIRD as competent and confident "barefoot solar engineers". By the time they returned, the building infrastructure at Rural Technology Park was taking shape for setting up the solar power plant on the 70 acred land site. Within no time the building for the power plant was ready and the procurement of the components was made for the installation of the 5 kilowatt power plant.

With their skill and expertise they installed the first ever 5 Kw solar power plant in the Technology Park. They installed the solar panels, the battery bank and set up the Rural Electronic Work Shop. On March 13th 2005 the Solar Power Plant was inaugurated by the ex-DG CAPART in the presence of Director General National Institute of Rural Development, Deputy Director General-NIRD and Dr.Manmohan Singh- Commissioner Tribal Welfare.

The 5 KW solar power plant can generate and supply enough power to electrify 15 houses in the rural technology park, which would illuminate 45 lights of 18 watts and 29 lights of 11 watts, 20 streetlights of 11 watts, in the Rural Building Center, besides computers, scanners, printers, coolers etc. in the Rural Electronic Work shop.

The BSEs also started making 12 volt 10 Watt solar lanterns. Nearly 100 lanterns were produced by them in 15 days of which 50 lanterns were bought by NIRD itself. 50 other lanterns were marketed to very prominent NGOs and government institutions. Since the past one year these lanterns were never brought for reference for any kind of servicing till date. Apart from producing the lanterns the ladies also provided after sales services.

2.2 Registration of the Society

During this time, sometime around May 23rd as the work of the BSEs was in progress a need was felt to register the society as the activities of the BSEs started expanding. Accordingly the bi-laws governing the functioning of the society were drafted, and a proposition for participatory selection of the governing body was put forward. Since the trained ladies are only four of them there was a need for inclusion of three others to make it into a body of seven to start the society, according to the Societies Registration Act. As there were only 4 members initially, 2 more members found loyal were incorporated from the community of women who were already selected from the Vijaynagar colony slums adjacent to NIRD campus. The association was formed with 8 members incorporating two Assistant Professors also to work/assist these ladies in the society until they become sustainable and manage the affairs of the society by themselves.

Accordingly Dr.Valentina and Dr.Madhuri were called in to work with the society and become a part of the society by working as executive members of the society.

A proposal for registration of the society with 8 members has been drafted incorporating the by - laws and was finalized for registration.

Three to four visits had to be made in order to complete the registration process of the society and receive the final registered certificate of the society. In the 1st visit information was sought on the procedures and documents required for registering the society from the registrar's office. In the 2nd visit the application forms were bought and all the required procedures were completed by incorporating all the signatures of the members of the governing body of the society. When the papers were submitted to the registrar's, he raised some issues like - clarification on payment of rent by the society to NIRD, as the solar power house is located in RTP- NIRD and the proposal of the association to work as an NGO and whether NIRD would like to charge some rent to the NGO for utilizing the solar house or give it free of cost.

So a 3rd visit had to be carried out after clarifying with the Director General on the issues raised by the Registrar. The Director General is not in favor of charging any rent to the solar house but the Deputy Director General was in favor of charging a minimum of 5-10 rupees rent per month. However, further clarifications of charging Rs. 5/- or if there are any minimum charges which can be charged had to be clarified from the Assistant Registrar (T) (NIRD) who referred to the records and explained that a minimum of Rs. 200 may be charged towards the utilization of infrastructure belonging to NIRD. This controversy put the file in a cold storage for one week. The DG was in favor of charging a nominal rent of Rs. 5 per month, and sought the opinion of DDG who took an assessment of the advantages and disadvantages of rent payment. The disadvantage of taking the rent was that, the DG has to provide an affidavit on behalf of the institute that they are charging a nominal rental from the society. While the advantage was a simplified procedure of declaration by the institute that no rental is being charged to the society for utilizing the solar infrastructure at RTP, NIRD.

After which the Director General insisted on completion of the procedures in whatever possible way that it was for the benefit of the society. Accordingly in the 3rd visit this clarification was made to the Registrar's office but the Registrar's office insisted on a letter from NIRD administration of its exemption of the rent to the society. Since the Director General was out of station it could not be produced immediately. So, the fourth visit was inevitable, in which the Director General's letter of exemption was submitted in the Registrar's office. Thus the society called women barefoot solar engineer's association (WBSEA) was registered on 9th June 2005 with the payment of a registration fee of Rs.200/- made to the registrar's office.

Accordingly the society was registered (details of membership in annexure -1) with eight members in the Ranga Reddy Registrar`s office as an NGO by the name WOMEN BAREFOOT SOLAR ENGINEES ASSOCIATION. Following the registration of the society it was decided that the society must have a logo for the official communication and business transactions. Accordingly a logo was designed for the society. The logo depicted the hands of an Indian woman holding a solar lantern, which was finalized by the BSEs through a participatory consensus.

2.3 Processing Bank Account for the Association

Opening an account in the State Bank of Hyderabad - NIRD branch in the name of women barefoot solar engineers association was the next event. The whole exercise took 2 days. The treasurer Yellamma and an executive member were given the signatory powers to withdraw the finances for carrying out the business. While the president of the association Ms.Chennamma and another Executive member were constituted as standby members to operate in the absence of any one member in the first set. Soon the account was opened and NIRD had deposited its first installment of 2 lakhs from the proposed Rs.5 lakhs as revolving fund to the association in order to enable the BSEs to start the enterprise.

2.4 Initiation of BSEs into the Business & Entrepreneurship

By this time the Solar house is full of activity and the BSE`s were engrossed in producing solar lanterns (10w and 12v). The ladies being trained are skilled in the technology but had a personality and adjustment problems as a result complaints started becoming visible as regards their attitude and behavior, to contain this problem an electronics graduate by name Mrs.Sevadan was deputed to take care of the solar house activities as she was a diploma holder in electronics. Mrs.Sevadan was fed up of the abusive slang language uttered by the BSE`s in the solar house and the unproductive time-spending behavior of the BSE`s during their lunch times. To rip off this behavior it was felt that there is a need for personality development sessions for the BSEs and educate them. Therefore many efforts were made by the executive member by spending time after the working hours to preach and teach moral lifestyle and explaining the need for the BSE`s to reorient their behavior to the new working atmosphere.

In August 2005 a meeting minutes register was opened as the Director General Shri Lalith Mathur opined that all the proceedings of the association should be recorded in the minutes book through consensus (after the meeting). As none of the BSEs could write, it was difficult to record the proceedings. Upon discussions it was found that `Zaheda' the Vice President of the association can write in Hindi. Though she is not fluent in writing word to word she had to be guided on entering the proceedings. However the association decided that she will continue to write the proceedings in the future meetings.

2.5 Discussions for Solar Electrification Project in Paderu

In the month of July, the commissioner, Tribal Welfare was approached to find out the status of agreements made on solar electrification project in Paderu. He clarified that the terms and conditions of extending the project to WBSEA have been finalized and the orders to take over the work can be picked up the next day. The next day, the team comprising of the young professionals and the WBSEA President visited the commissioner's office and received the orders to launch the solar electrification project in the selected villages and was presented before the Director General -NIRD. DG congratulated the team and asked the team to gear up for the major responsibility ahead.

3

Action Research Interventions on Solar Electrification Project

3. (A) Introduction on Action Research in PADERU

Two major Action Research projects on solar electrification in the tribal villages of the country were conducted by NIRD Women Barefoot Solar Engineers. The first one was initiated in the state of Andhra Pradesh, Vishakhapatnam District, Paderu Mandal, chintapalli Village; while the second project was launched in the state of Orissa, Koraput District, Tingnaput village. This chapter discusses all the action research interventions and events that led to the initiation of the solar electrification project in the study villages. The sequence of events on how the community-managed rural energy systems and powerhouses that work without dependency on exogenous skills and technicians, demonstrate that the communities can do it, and exhibit solar electrification as a viable and cost-effective alternative. In a move to establish this objective, illiterate women trainees were given the skill into their hands to solar electrify the households and also manage it.

In August 2005 the Tribal Welfare Department sanctioned the project on solar electrification of two un-electrified villages in Vizag District- Thamingula and Pusalapalem villages of Paderu panchayat, Chintapalli Mandal, Visakhapatnam district. The villages were identified by the Tribal Welfare Department and the criteria for selection of these villages is that they are remote, backward and inaccessible and located in the naxalite area. Thamingula has 88 houses and Poosalapalem has 40 houses. They are both between seven to nine km from the main road and the nearest electrified village. They pay Rs 15-Rs 20/liter in the black market for procurement of kerosene and have to walk 9 km down the hill to reach the PDS shop.

After assessing the energy needs and demands of the village, the villagers consent was sought on electrification of the village. The villagers responded affirmatively and a demonstration of the working of the solar lantern was made. A detailed understanding of the solar energy system proposed to be set up in the villages was provided to the tribal households through discussions and demonstrations. An over view on the socio economic status of the action research villages gives a further understanding on the interventions required in the villages.

3.1 (A) Socio -Economic features of Bhagatha Tribe:

Thamingula is a small village comprising of 88 households. The village is inhabited by tribals belonging to Bhagatha sub tribe while poosalapalem village is inhabited by valmiki sub tribe.

The bagatha, also called bhakta, bhagada, bogatha and bhagatha are scheduled tribe inhabiting the hill tract of Visakhapatnam district (Thurston, 1909). They are concentrated in the mountain valleys of Paderu, Sujanakota, Karakapalli, Tajangi, Devarapalli and Peddavalasa which comprise two-thirds of Visakhapatnam agency.

Being expert archers, the Bagatha served the 17th Century Golconda and gangaraju Madugole chieftains of Visakhapatnam Agency as their army personnel. They showed so much devotion that they were recognized as their devotees or bhaktas. The name 'Bhaktas' later became Bagathas (Thurston 1909; Reddy 1979). In appreciation of their military services, the chieftains appointed them as local chiefs, muttadar of the hill tracts. Due to this shift in status, the Bhagatha claimed themselves as a warrior tribe, expanded geographically, developed politically and acquired social supremacy in the tribal area over a time span of about 50 years from the late 17th Century to the early 18th Century (Reddy, 1972).

The Bagatha are well-built, skin color varying from light brown to dark brown. Their hair is coarse, straight and black. They have predominantly dark brown eye colour. The Bagatha speak Telugu with their hill accent. Bhagathas residing at the marginal areas of Orissa State speak a corrupt form of Oriya. Bhagathas are simple and traditional in their clothing. Women wear simple sari (cheera) where in one end is tied around the waist and the other end (paita) carried to the right shoulder (across the shoulder). The sari tied in this fashion hangs down only up to the knees. The women do not wear a blouse. Bagatha women wear a number of modern ornaments like nose-pins, ear-studs, anklets and ear-pins made of brass or bronze.

The staple foods of the Bagathas vary. They consume different types of millets and rice. The millets are converted into flour and cooked as gruel called ganji, ambali or thompa or dumpling -sangati. These foods are supplemented by pulses, and a variety of vegetables. As the cereals exhaust they take foods like edible leaves, flowers, fruits, seeds, mushrooms, birds, mammals, fish and insects which are available in the forest. The Bagathas are non-vegetarians and they relish field rats while Beef and pork are a taboo for consumption. Men, women and children smoke beedis and take intoxicating drinks.

Bagathas usually are Nuclear families. The societies are patriarchal which bestows a lower status to the wife in the family. The sons live with their families in the village after marriage but the daughters move away to their husband's houses after marriage.

The principle of agnation is practiced in the village where daughters do not possess any right to inherit the parents property .The girls in the community are given in Marriage within few years of puberty between 13 and 16 years of age, and boys between 16 and 20 years of age. Marriages are contracted outside the lineage, surname group and clan but within the tribe. In these societies also the modern dowry system is replacing the traditional bride-price which involved the exchange of oil. The sources of Bagatha's income depend on agriculture, shifting cultivation, hunting, fishing, domestication of animals, trade and barter, occasional labor for wages, and self-employment. Agriculture is the major source of livelihood. Shifting cultivation (podu) is usually practiced by them.

Apart from individual and communal hunting methods, food gathering also constitutes an important livelihood intervention. The fibers of different trees are collected for making ropes , and the leaves of gigantic camel's foot (adda) is utilized for making leaf plates, while the soap pods (sheekai) myrobalams, bixa fruits (jabra), mohua fruits, medicinal herbs, honey and brooms are collected and sold in the weekly markets.

In the system of social hierarchy of the tribals in the Visakhapatnam Agency, the Bagatha have the highest place. The Bagathas are dominant landowning and food grains producing tribe.

The Bagathas have availed the educational facilities. In health they have accepted the family planning programme.

3.2 (A) Socio-Economic conditions of Valmiki Tribe

The Valmiki tribal groups are concentrated in the agency area of HYDERABAD, Srikakulam.Vizianagaram, East Godavari and Vishakhapatnam districts. According to Thurston valmikis claim a descent from Valmiki the author of Ramayana. The valmikis are notified as Scheduled Tribes in the agency areas. The valmikis are also called as Paidi or Boyas in some areas. The ancestors of the Bhagatha community are considered to be the Mals of the plains. The valmikis are predominantly rural where more than 98% of the populations live in the rural areas.

The valmikis (men) wear Gocha, a shirt and a baniyan. The women usually wear sari which is wrapped around the body and tied up on the right shoulder, tattooing on the fore head and the arms is a common tradition amongst the men and women of the community. According to 1991 Census, their population is 55,836.

The total literacy rate among Valmiki is 39.87.The Valmiki tribe is divided into following Gotrams in order to regulate the institution of marriage among them in Visakhapatnam tribal areas. Naga Bowse (snake), Matsya Bowse (fish), Pangi Bowse (kite), Jilla Bowse (goat). But these clan names are absent in tribal areas of East Godavari district. Marriages by mutual consent, marriage by elopement, are the methods of acquiring mates, widow remarriages and divorce is also permissible in the community.

Valmikis are agriculturists and forest laborers. Some of them became traders and petty moneylenders. They sell the earthen pots also in the shanties. They practice podu cultivation on the slopes of hills.

3.3 (A) Features of the Village

All the women of the village are members of self-help groups. The SHG movement in the village is strong which started in the year 2002. There are 7 SHG groups in the village. The SHG's are federated into Mandal Mahila Samakhya which is initiated by Velugu project. They usually hold a meeting once in a month and collect rs.30 per month. The accounts are maintained by the members of the groups themselves.

Nearly 90% of the women are uneducated and only 3 -4 women are educated up to class five.

More than 50% of the households are landless. Agriculture is the main livelihood activity and most of them work on their own fields while the landless laborers work in others agriculture lands as laborers. The wage remuneration for employment is Rs. 20 per day in the agriculture season. Under velugu project rs.1, 88,022 have been given as micro credit to meet agriculture and economic needs of the Tribal households. The credit is provided at Rs 2 percent interest rate, which the households have utilized for agriculture purposes.

During off-season, livelihood is sought through employment in off farm activities. The secondary activity of the Tribal households is collection of tendu leaf from the forest. Each bundle of Tendu leaves comprising of 1000 leaves is called as a Kim. As one Kim of Tendu leaves are collected they are sold out in the market. In a week, nearly 1 Kim of Tendu leaves are collected and sold in the market at the rate of Rs. 60 /Kim. Tendu leaves are available on a sustainable basis and in a bulk in the village; a technology intervention in leaf plate making can enhance their income. But as on date no such efforts are seen in the village but the households have shown their keen interest in seeking training on leaf plate making.

As far as the community infrastructure in the village is concerned there is one Anganwadi center , a panchayt house, and a community hall in the village .There is also a small multipurpose store in the village which is able to supply rice, oil wheat and essential commodities for the whole village.

3.4 (A) Previous Electrification Attempts in the Village

The village was provided with electrification in the year 1995 under the rural village electrification programme of the then Chief Minister Mr.N.T.Rama Rao. At the entrance of the village an inaccessible road of 7 Km (constructed out of WBM (water bound Macadam) is bordered by eroded and uprooted stand-alone street lights, which tell stories of previous instances of electrification wherein the government seems to have spent cores of rupees in providing this electrification to the village. Apart from street lighting each village has the skeletons of solar street light systems in the interior parts of the village which gives an impression that it must have costed the government's exchequer heavily in facilitating these systems;. A discussion from the villagers gave us to understand that the systems were functional for a brief period of 2-3 months and one fine day the batteries blasted (without any local casualties), and the lights became nonfunctional since that day. When the WBSEA team visited the village the Skeletal remains of these lighting systems were standing out as scarecrows for the village. But the panels were dismounted by the tribal youth to run the music system, where in the tribal families exchange the panels to run the music systems as and when required.

3.5 (A) Village Infrastructure

The Primary Health Centre is located at a distance of 11 km from the village. The village has been declared as the malaria sector which mandated the availability of an Health Assistant and an ANM round the clock but unsurprisingly they are not available in the village.

The long stretch of 7 km road leading into the village is a WBM based road. With heavy rains there is serious erosion of the soil exposing the stones on the road which makes the terrain difficult and in accessible to the village. This stands out as a hurdle in initiating either a developmental activity in the village or reaching out any facility for the tribal household's .This may be one of the reasons for the village to remain backward, despite serious interventions being made to develop the villages by the government. However the SGRY works like roads and wells seem to have taken up in the village in a limited way.

Fig. Solar Electrification Being Provided a Decade age.

3.6 (A) Pilot Study Visit to Paderu

On 10th August 05, the WBSEA team comprising of the president, the Vice President, Technical assistant Mr. Satyanarayan, the young professional and the executive member left for Paderu to select four trainees for the project sanctioned by the Government of Andhra Pradesh. The administration has released Rs. 10,000/- as the revolving fund into the WBSEA account, to meet the expenses on the field.

As soon as the bus took off, Chennamma the president who is not used to long distance traveling started vomiting in the bus. But she managed to take care of herself very well without soiling herself or the place where she was sitting until the next morning till 9 am.

The distance from Vishakhapatnam to Hyderabad is 600 km and from there on Paderu is 40 km which nearly takes 4 hours to reach Paderu mandal.

The road leading to Paderu mandal is a ghat road and the bus has to follow serpentine rounds around the mountain. The road has deadly sudden and deep hairpin bends with a narrow BT road of two meters width which leaves no space for a vehicle either to overtake nor give space to a vehicle approaching in the opposite direction. All throughout the road, the journey is interrupted by few limited stops. However the scenic beauty is breath taking, with the sky seeming to be so low as that it can be touched. With the cloudy weather the bus disappearing and reappearing occasionally in and out of the clouds was magical to watch. But a peep from the bus gave pangs of fear as the bus took centripetal turns around the mountain. No single human being can be spotted on the road; however trees stood tall stout and strong bordering the roads. Every bend and turn was skillfully driven by the driver in a high speed.

Paderu is 3000 mt above the sea level. Here the forests are bestowed with fertile lands and are endowed with rich vegetation. The local areas have rich coffee plantations as the land is suitable for coffee production. In order to encourage coffee plantation and production the government has set up a coffee board in the mandal which conducts capacity building trainings to the farmers in coffee plantations and production.

The WBSEA team reached Paderu by 3 pm. With the onset of monsoon it was drizzling all through the day. The welcome scene as the team alighted the bus in Paderu seemed as though we were in Ooty. The team located the ITDA office and proceeded to meet the Commissioner but disappointedly the Commissioner was on a field visit in the local area as the whole village was reeling under the malaria attack and malarial deaths in the village.

Despite information and official communication of the arrival of the team in the village for solar electrification project there was no one to provide any kind of accommodation. The Paderu ITDA office failed to make the basic arrangements for us when we arrived.

Disappointedly the team approached the commissioner's PA, who arrived after his usual lunch time in the office and the accommodation was provided for the team at the Primary Resource Centre.

The Primary Resource Centre is a training centre managed by VELUGU - an NGO sponsored by the Government of Andhra Pradesh. But to the utter dismay of the team, the training centre was also full and lacked any accommodation except for a dormitory and a single VIP room. The VIP room was provided to the Executive member. But as there are two other women members of WBSEA, the VIP room was given to the technical assistant and the young professional and all the three ladies adjusted in the dormitory.

The dormitory has a bathroom at a distance, but it lacked sanitation in all respects. Its maintenance in terms of sanitation was totally negligent and a peep into such a sanitary outlet makes one imagine the importance that is given to hygiene and sanitation in the rural areas. However there was an important need to get freshened up .Overlooking the status of the toilets the team refreshed and proceeded to meet the velugu officials who were supposed to accompany us into the selected villages.

On that day only one official Mr.Srinivas Baba was working as Assistant Project Manager in the velugu office. After a detailed discussion of the project the official agreed to arrange a vehicle for field visit to Thamingula and Poosalapalem villages.

The next morning i.e. 11th July, the team proceeded to Thamingula village and the team was accompanied by 2 community coordinators Shyam Sunder and Mahesh from Chintapalli, the APM Mr. Krishna Rao. The whole team of 10 members reached Thamingula by 10 am.

On reaching the village the Sarpanch Sundru Bangariah, the Vice president Mr.Bangaru Babu of the Panchyat and SHG Federation leaders S.Papayamma, P.Laxmamma and T.Kondamma welcomed the WBSEA team.

3.7 (A) Visits to Thamingula Village - First Proposed Village for Electrification

Thamingula was the first village taken up for the field work. As soon as the jeep crept into the village, children were the first to receive us as spectators. From there on, the women of the village started gathering one by one, followed by the youth and men of the village.

By the time the team reached the village a political rumor was in circulation by opposition Telugu Desam party which made allegation against the ruling congress party that hundreds of people are dying every day in the village due to Malaria fever and the ruling party has failed to cater to the welfare of the Tribals. Consequent to the uproar, the medical officers and village health workers were pursued by the collector to provide anti-Malaria tablets and visit and attend any person suffering from ordinary fever also. Though the crowd was practitioner (medical) centric as we entered the village, soon we became the centre of attraction, and enquiries started pouring in to find out who, what, when, where and why we arrived into the village.

The APM requested one of the enquiring ladies to call and gather all the SHG women in the village and Assistant immediately organized the meeting with all the villagers by first calling the leaders of self-help groups. Within minutes, all the women of the village gathered, followed by the men, young and old. All the villagers gathered at the peepal tree where the panchayat usually meets. The WBSEA team introduced themselves and a rapport was built with the villagers by discussing the attack of recent epidemics and the medical attention given by the government.

A discussion was initiated by the WBSEA team on the PDS system in the village and the availability of kerosene, its cost price per liter, the access, experiences on cheating on the content of kerosene provided by the PDS, the advantages and disadvantages of its usage. The questions so posed to the households in the village retrieved the saga of toil and deception undergone by the village community in procuring the allotted ration.

Systematically the purpose of the visit was explained to the localities. There was pin drop silence as the introduction was going on. Observing their attention and interest in the discussion the president of the association posed a question to the villagers - do you want lights in your village? The villagers looked at each other but refused to comment. The women were very shy and squatted on the floor covering their whole self and the little ones hanging out in her lap.

The children looked very poor wearing worn out and fettered clothes with a perennial mucous flowing out from their noses. Small boys (some) wore extra-large knickers which testify that the clothes are not theirs. The little girls flocking around, carrying their younger siblings though they were unable to do so, was a pathetic sight. A small child unable to carry herself is volunteering to carry her siblings. Of course the men were totally boozed and stinked all the time that they spoke. In general the population has an average health and are not starved of food and health.

The tribals though shy and refused to talk, some of them especially those that are in contact with the towns, have all the knowledge of ongoing rural development programmes. When a survey was made, all seem to have given us a wrong economic status being apprehensive that either they may be asked to pay some or a part of the money towards the programme or they may not get the system which may prove their above the poverty level status. Most of them possessed lands, almost 70% of them and each family possessed not less than 2 acres of land. The households were as such self-sufficient.

As the team from NIRD demonstrated the solar home lighting system and the lanterns, the villagers became curious and every one of them wanted to operate the lantern and see how it functions. The team has also carried some additional spare parts to explain the function and the delicacy involved in assembling the technology. Men took all the parts into their hands and examined them and it was very surprising to learn that these tribal youth were well equipped with the technicalities of the electrical works. Five to six households in the village also possessed the transistors with which they heard the the news bulletin and this surprised the NIRD team. This infers a paradigm shift in the culture of the people and westernization and industrialization seem to have penetrated well into the hilly tribal households also.

Later it was learnt that in the NTR regime, the village was considered for solar electrification by providing the street lights. Within 3 months of its installation the lamps became defunct and the village youth dismounted the panels and connected them to their transistors to run them even without batteries.

As the parts exchanged the hands, men were convinced that the technology is not that easy but requires some inputs through training to gain mastery over it. However, the women were not convinced to participate in the transfer of technology, may be because of the lack of knowledge in the technology.

But to the utter surprise of the NIRD team, the women silently took the solar lantern into a house, closed its windows and doors and switched the lantern 'on' and observed the effectiveness and illumination of the lantern in the room and only after being satisfied they opened the doors and came up to us and expressed their willingness to be a part of the transfer of technology.

After the consent was given by the villagers from Thamingula village, we put forward the proposal of taking 2 unmarried girls for training at NIRD for six months. This issue had to be put forward very effectively and systematically since the tribals never send their girls out of their village. Therefore discussions had to be evolved in such a way so as to make them understand by delving on the previous experiences of electrification through solar power and grid power and asking them questions on when it was installed and how long they functioned, who could repair it and what are the consequences of not getting the systems repaired. The tribals were very understanding and as they were being questioned one after the other, they replied, emphatically and optimistically.

After such a detailed endeavor, the team asked the villagers to nominate the names of two girls, who were intelligent and willing to come to NIRD. All the women flocked together and started discussing, without bothering about us. In the meantime as men were ideally watching their women in a process, the team made a selection of the member for the constitution of village energy and environment committees. After a break of half an hour the NIRD team approached the women's group which was seriously involved in discussions. By now the women were ready with the names of two girls that will be sent to NIRD for training and also the names of the women that will be the participants in the village energy and environment committee.

3.8 (A) Kerosene Procurement in the Village from PDS

The Tribal households are entitled to get 3 liters of kerosene from the GCC. But here also the Tribal households are deceived by the retail agent who distributes the ration. Usually the measure of one liter is filled with a deft hand and the quantity so procured is not complete and the cost is high by Rs 2-3 per liter from the market rate. Sometimes despite waiting the whole day the households may not be able to procure even those 3 liters of kerosene which they are entitled to at the rate of Rs.10 per liter by the Girijan cooperative society. Sometimes the tribal households purchase kerosene at Rs.20 per liter. On an average a household consumes 10-11 liters of kerosene per month. The additional requirement is met through purchase in black, all through the year which is attained @ Rs.15-20 /liter. Each household illuminates nearly 2 lamps per day.

3.9 (A) Initiation of Solar Electrification Project; People's Participation for Sustainable Solar Electrification Programme

Sustainability of a programme in rural development has been a major hurdle since the inception of developmental programmes in the country. Application of any technology for the community was not individualized but was perceived as a public asset. Such an attitude in the beneficiary withdraws the service providing ability and once the systems fail to function there is a laxed attitude to attend to the repairs. and the whole effort ultimately becomes a fiasco. To overcome such an instance the following measures are taken:

- Formation of Village Energy and Environment Committees
- Peoples contribution of an installation amount of Rs.1000 per household at the time of installation
- Collection of user charges @ Rs.100 per month for utilizing the electricity.

When these i.e. ever nere placed before the community, all the members of the village agreed to pay an amount of Rs. 1,000 at the time of installation of the Solar Home Lighting System in their respective households. The villagers have requested the WBSEA team to collect the amount from the households in the month of December as they expect the returns from harvest during the time. The villagers have also agreed to pay an amount of Rs. 100 every month after the installation of the home lighting system. However they have agreed to pay this amount in lump sum in the agriculture season if they fail to pay it in the off-season times.

3.10 (A) Selection of Trainees

As per the project plan, two trainees were to be selected from each village. But before selection, getting the consent of the village to send them to NIRD was little ambiguous as the tribals do not go out of their villages and are very shy. Against the apprehension of whether they will send their girls for training WBSEA assured the village that the girls will be taken for the training and WBSEA is completely responsible for their welfare. After a brief consultation the village agreed to send the girls which was further reinforced by the sarpanch that if the village has to prosper, then the trainees will have to be sent for the training. As per WBSEA's opinion the following criteria was adopted for selection of the trainees:

1. To select women members only
2. She should be married
3. She should not be in the family way
4. She should be able to stay at NIRD for 6 months
5. She should have minimum education up to 6th class.

Against these set demands it was very difficult to select the trainees as all of them could not satisfy all our demands. If one had education, she lacked a family status, while if the other qualified in other aspect their husbands were reluctant to send them for the training at NIRD. However overcoming all the hurdles, WBSEA could zero in on two girls for training at NIRD who were nominated and finalized by the village. They are Sunna Papaiamma W/o Krishna and Pravilika W/o Ramesh. The two girls are co-sisters from a single family. Though the girls looked terrified, we assured them that they will be taken care of by WBSEA.

3.11 (A) Village Energy and Environment Committee

A village energy and environment committee (VEEC) was constituted with the following objectives:

1. To monitor all the energy needs and demands of the village
2. Provide security against misuse and thefts of the systems
3. And work for the sustainability of the systems
4. Collect the user charges and get them deposited in the bank
5. Collect the installation charges and deposit them in the bank
6. Take collective participatory decisions in case of replacing the systems (if need be) against theft or damage of the systems.

A VEEC committee comprising of 20 members with sarpanch of the village and members of SHGs was constituted (Details in Anexure-1)

The committee has nominated Pottur Rajababu and Kondamma and the Chairman of WBSEA-Chenamma, as signatories to hold the joint account. The monthly collection of rs.100 is the responsibility of the committee (VEEC) and the committee itself will deposit the money in the bank, which is opened on their account, and utilize it for the purpose of repair, maintenance and battery replacement in the future. An amount of Rs.1000 so collected at the time of installation of the solar home lighting system is also deposited in a fixed deposit.

3.12 (A) Visit to Poosalapalem-second Proposed Village for Solar Electrification

After completion of the initial demonstration at Thamingula village the team left for the second village called Poosalapalem. This village is 2 to 2.5 km away from Thamingula village. The villagers in this village meet their livelihoods by working in the agriculture fields (nearly 50 % of the households have agriculture lands).

They grow Rice, Ragi, Turmeric and Lentils and usually are involved in monocropping and subsistence economy wherein all the produce is utilized for consumption by the family only. During the past 10 years, the village received paddy seed from the government at the time of sowing.

The agriculture produce is not marketed while the cash crops are marketed. On an average each person gets an employment opportunity for one week in a month .The adda leaves and other forest produce collected like tamarind, soap nuts etc collected from the forest are marketed in the weekly shanty, and this constitutes their secondary livelihood option through which they earn an amount of Rs.150-200 in a week.

There is one primary school for the whole village. And only 8 people are educated in the village up to 10th class. People are more traditional in their attire in this village, which is tribal attire. The women are very shy and docile and usually take alcohol.

The SHG activity was started in the village in the year 2000; however the SHG activity is very weak. There are 3 SHG's in the village. The women are simply crediting the money in the banks and they are not aware of the thrift and credit activity of the SHG, the banking facilities, the subsidy and other privileges given by the government and the NGOs. The women also narrated instances where in the money collected by the SHGs was swindled by the sarpanch and government officials who promised to deposit the amount in the bank. The accounts are maintained by the velugu facilitators. The SHG's meet once in a month and collect and deposit rs. 30 per person. SHG's are bank linked and they have an access to the bank situated at Chintapalli. It costs rs.22 for the to and fro fare when the women go to deposit the SHG money in the bank.

After completing the pilot visit to thamingula village, the WBSEA team left for the poosalapalem village. The poosalapalem village is 4 km towards the north of Thamingula village. Only 1km is motorable while one has to walk the rest of the 3 km. To reach the village. It was almost 3 pm by the time the team left. There was not food in the afternoon however the WBSEA team did not complain. After a 45 minute walk through a partial desert, the team reached Poosalapalem village. Every person in the team was gasping for breath as the village was located at the top of the hill.

The hill was on a 60 degree gradient and extra pressure had to be exerted on the earth to reach the poosalapalem village. Additionally, carrying the solar lanterns and panels for demonstration to the top of the hill made our job even tougher. The path leading to the village is through a forest. Many small rivulets run across the path. The path is undulating and rocky and there is no beaten path taking us to the village. It was difficult for the WBSEA to manavour through the rocky mountain to reach the village. It is a Kutcha road with steep edges. The whole village is mountainous and the households live scattered all over the mountain. The village has 35 households .The houses are semi pukka with tiled roofing. The little girls in the village do all the job of cleaning the dishes and getting water for the entire family from the closest stream. The stunning scene of a 14 year old girl carrying 6 vessels one on top of the other placed on her head (cleaned to a sparkle), filled with water and walking on the rough terrain surprised the WBSEA and made us understand of the toil that a girl child undergoes in the village. It completely amazed the team. Latter the girl was interviewed and pictures were taken. It was learnt from the discussion that a girl child is made to perform these tasks ever since the age of 5 and today she has no problem carrying 6 vessels full of water over her head.

As the team was climbing the hill it came across two small water rivulets flowing across, from the top of the mountain; the water was clear and the speed of the water flow was slow and constant. This small stream culminates at a point through the rocks where the pressure of the water is 2 inches and it is learnt that this water flow is perennial. This is a small spring from which the villagers fill the water and utilize it for drinking purposes.

When the team visited the place, the place was full of girls in the age group of 5-14 years. Each one was busy scrubbing the utensils and cleaning them. All the girls sat closer to each other and were scribbling out the carbon from the vessels to a sparkle and were lost in conversation with each other. The source from where the water trickled was made conducive for the water to flow by arranging a small cement sheet. The water source was bordered by fungus yet the drinking water is collected from the same point. This seems to be one of the reasons for the outbreak of fevers in the whole village especially in the rainy season, which cost the villagers their dear fortune for the season.

As soon as the team reached the top of the village, as usual the children were the first ones followed by women and here we were just treated us as though we were their own relatives. The SHG members immediately organized seating and tea arrangements for us. Simple information to call all the villagers was given to an SHG member and the information as regards a quick meeting was broadcasted through an echo made by the tribals standing on top of the mountain. In no time all the households that that were available at home arrived to participate in the meeting. Similar to the demonstration made at Thamingula village the Poosalapalem village was also provided with the same demonstration.

Here the tribals were very docile and did not interact much and we had to speak to them for longer hours to extract their ideas on solar electrification. However, we were struck here in the selection of the trainees as we could not find even a single girl who was educated. All our enthusiasm died down and fear gripped us and we were in search of alternatives on - What if we cannot get the right candidate for the training? The girls were either old or too young and unmarried and not educated at all.

We were greatly disappointed and did not know how to handle the crisis. Some of the SHG members were quick to remind others that some of the girls are in the forest and that they have gone to collect Adda leaves and wood for cooking and could be expected any moment. The team had to wait till 7 pm, when the girls arrived, but the whole batch did not arrive. It was getting further delayed and the velugu officials accompanying NIRD team were compelling us to leave the place immediately, fearing the naxal attacks. Thus a decision was taken by the team to visit the village again the next day.

The team boarded the jeep and the vehicle could run only a distance of 10 km of the destined 40 km from Paderu. After which the vehicle failed to move further as it was an old vehicle and there was serious accumulation of dust in the engine and diesel filter of the vehicle, as the vehicle had to waddle through loose mud on its way, in addition the vehicle exhausted the entire diesel. The young professional and Satyanarayan had to frequently get down and pump the filter by lifting the bonnet of the jeep, which enabled the jeep to move 2 meters only, after which it would stop. This exercise continued for 2 hours on the road. When the boys were struggling on the jeep, many other vehicles passing by also stopped to enquire the reason for failure of the vehicle. Whenever such vehicles were encountered the only fear flashed in our minds that they must be the naxalites. But all efforts to move the vehicle went vain.

The driver informed us that we were caught up in the middle of the forest, under a big mountain and a very prominent adda of the naxalites and we will have to sleep in the vehicle until morning and there was no option. The insect's chirping noise was the only noise heard around. By then it was 9 pm and anticipation for the last bus was the only hope but the last bus also seems to have left and there was no hope for a bus until dawn. Then WBSEA team decided to lock the vehicle on the road and leave by which ever transport that will be available. But till 6 am there was no public transport available.

During the stranded time on the road the driver started telling us the tales of naxal attack on the people in that particular village which took place the previous day in which one person was killed. With this all the courage of all the members was sapped and became volatile and each minute wait in the vehicle made the team psychedelic. The situation scared all the members of the team. Some vehicles going the same way stopped and tried to help us but all in vain. The members in the jeep were waiting for it to be dawn. The whole night the team spent hallucinating in the vehicle.

By 6.30 am the noise of the first bus alerted the team and as it stopped we boarded the bus and went to Paderu. As we had already promised the Tribals in Poosalapalem that the selection of the trainees will be finalized the next day, the WBSEA team had to rush back immediately to poosalapalem village in the morning. The members freshened up quickly and immediately hired a commander jeep and left for Poosalapalem village.

The running time from Paderu to Thamingula was 1 hour. But to our utter surprise, the jeep started climbing new hills and it was more than an hour that the jeep was running and the dust on the roads made it difficult for the team to recognize the place. Sensing that we are not on our intended destiny to Thamingula village, the team stopped and enquired from the Pedestrians. Then we realized that we have traveled 15 km away from Thamingula village of Chintapalli mandal but were heading towards Thamingula village in another mandal. Irked by the misunderstanding, the driver was taken to task by the team, who sped fast and took us to the village in another half an hour.

By the time we reached Thamingula village of Chintapalli mandal it was 12 pm and further as we climbed the top of Poosalapalem village it was 12.45 pm. To our dismay not a single person was available in the village to give us any information. After waiting for a while we learnt that the tribals had waited for us from 8 am to 10 am and after that they had left for the church. In Poosalapalem village a majority of the population are Christians. The team had to wait till 1.30 pm when all the tribals returned from the church and gathered for discussions. It was here that all the girls of the village came forward to attend the meeting. But surprisingly by the time the team reached the village the villagers had already decided the names of girls that would go for training at NIRD. The village nominated Pangi Kumari w/o Chitti Babu - she is educated up to class 2nd and Sagina Santhoshamma w/o Chinnabbai who is also educated up to class 2. Here WBSEA had to settle with little lesser educated girls, since there are no educated girls in the village.The committee also nominated two members to operate the joint account with the signatories Chitti Babu and Sagina Laxmi (details of the committee in Annexure 3).

The WBSEA team took the selected girls to Thamingula to introduce the other two selected trainees from Thamingula village so that all 4 of them can know each other, discuss and chalk out a programme to reach NIRD for training. It was here the president and the vice-president gave their assurance to the localities that they will be well taken care of at NIRD. The president was more cooperative and spoke at a closer proximity with the villagers and made the tribals feel at home. Then the president gave Rs. 2000 to meet the travel requirements of the team. The Velugu official assured us to send one of the community coordinator to accompany the girls to NIRD. In total satisfaction, the team left for Paderu. As the team was leaving, the whole village walked down the hill to give us a sendoff. With this, a long term relationship was built with the villagers.

3.13 (A) Meet with the Project Director

After the completion of the preliminary works in the project and formation of the committees in the villages, WBSEA team met the PROJECT DIRECTOR Shri. Siddhartha Jain PD, ITDA, Paderu and briefed him on all the activities that have been carried out by WBSEA in the village. The Project Director congratulated the team for completing the first phase of work successfully and gave his observations on the Solar electrification based on his previous experiences and the PD embarked on the issues of sustainability of the systems in the project site as no one took the responsibility to maintain the system as it was seen as a public asset. However WBSEA team assured him that the village environment committee was formed with an idea to oversee such short falls and sustain the project. Further the PO insisted that the intervention if ITDA may be concealed with the people in the village as a precautionary to curtail dependency in the people for maintenance and other reasons. However he was of the view that the programme should be launched in such a way that the villagers themselves should take the responsibility for the maintenance and sustainability of the systems in the future. He assured WBSEA in extending his total cooperation and support of any kind in meeting the project objectives. He also asked WBSEA to send a list of the members of the committee so that his office will be able to monitor the activities of the committee. The PO also emphasized that only after being convinced that the villagers will be able to handle the solar home lighting system on a sustainable basis that he will be able to grant permission for the implementation of the phase -II of the project involving street lighting systems in the village.

The 4 selected barefoot solar engineers selected from the two villages left Paderu on 20th of august 2005 for 6 months training at NIRD which was started on the 22nd of August 2005.

3.14 (A) Departure of WBSEA team from Paderu

The WBSEA team decided to get back to Hyderabad with the successful completion of all pilot initiative of the project and started off to Hyderabad in good faith, that we'll certainly get some conveyance either a bus or a train taking us back to Hyderabad. We started from Paderu and reached Vishakhapatnam bus stand by 8.45 pm, but to our disappointment, the last bus had already left by 7.30 pm. We were seriously struck as there was no way out to take a vehicle to Hyderabad. The vice president (Zahida) was too tired and was dropping off to sleep at the bus stand itself. There was no option but to wait until the next evening and board the bus to Hyderabad. The tickets were booked for the next day and the young professional (NIRD) and the technical Assistant went in search of hotels but there was no hotel which could accommodate us in our budget (in the range of Rs. 150-200) in the area. The only option was to shell down more money and take accommodation in a hotel. In a puzzled condition as the WBSEA team was contemplating on the next move, we came across a friend who promised to get us an accommodation in YMCA.

We waited till 10.30 pm but there was no information, but by 1 pm the friend had arranged us a big hall and food for everybody. Soon we took our luggage and went to the new accommodation, freshened up and happily had our dinner and went to sleep, all the ten of us putting our beds across the hall where ever we were comfortable. After a satisfying sleep the team decided to go to the seashore in the morning and spend some time at the seashore as there was time to leave for Hyderabad. By 9 am, the RTP consultant, NIRD, Brig. Reddy called us on the telephone and took information of our activity and congratulated us on the efforts made and conveyed it to the Director General Shri Lalith Mathur. On 16.8.05 early hours the team reached Hyderabad and decided to take an off for the day as all were tired.

3.15 (A) Budget Finalization for the Solar Electrification Project

To start the implementation of the project WBSEA Team had to work on the budget, the Completion of the pilot study gave us ample knowledge and understanding on financial planning for the project. Accordingly the budget was drafted in association with the technical assistant from Social Work Research Centre (SWRC) who followed the guidelines of the previous projects and finalized the budget for the project. It was submitted to the Director General for final approval and onward submission to the commissioner Tribal welfare Department. But the DG was not happy with the costing provided as he felt that the cost was too high. The file was sent back to us and revisions incorporating changes suggested by DG were asked to be carried out. Again the whole budget was reworked. The team had to work till very late in the nights to complete the work, however with the cooperation of the technical assistant the draft used to be completed.

This way the financial statement had to be reworked and resubmitted for 5 times as we were novices to the field, then the ultimate budget on its 5th time of submission was approved by the DG and a copy of the budget was sent to the commissioner Tribal welfare Department. The commissioner did not take much time but quickly perused the details and approved the budget. Within one week the required finances to carry out the project were released and we could put the next step in taking the project forward.

In the approval of the project, the first phase of the project involved allocation of financial resources for of 132 home lighting systems in Thamingula and poosalapalem villages, put together, training for 4 girls selected through participatory method for 6 months at NIRD, training for Village Energy and Environment Committee. The second phase of the project involved provision for installation of a 1 KW power plant through which 10-12 street lights, Community TVs in both the villages and a water distillation plant will be provided in the village. The total cost of phase 1 of the project is Rs.20.25, 094 and the total cost of the phase 11 of the project is Rs.5, 15,751.00. After the completion of phase -1 of the project and based on an impact assessment study report and sustainability of the systems and the project the phase -2 of the project is to be initiated.

3.16 (A) Management of Solar Technologies and Women Empowerment

On 21.8.05 the four trainee girls selected for the training arrived at NIRD. The accommodation for their the stay was made at Quarter

No: 2 in NIRD campus. The young professional and the president WBSEA purchased all the essentials for the stay of the trainees like the beds, utensils, groceries etc. as they were to cook their own food and go for the training. But as the finances were not released by A.P. TRIPCO for their maintenance, the girls were accommodated in guest house - 3 temporarily. From 22.8.05 the training commenced for the tribal girls at RTP, NIRD.

The first few days, the girls were comfortable and happy as they were exposed to a new environment. For the first time in their life they were eating food tasting deserts and fruits and living in a completely different lifestyle. The rooms were facilitated by fans, television and bathroom with warn and cold water supply. They loved watching movies as they retired from the training. The food at the guest house was palatable for them. Gradually they started understanding the work and the technology. Of all the 4 girls, Ms. Papayamma was educated up to 5th class and she was seen to be grasping the intricacies and nitty gritty of the technology very effectively and she was also fast in her work performance. Recognizing her talent, she was asked to assist and help the other three trainees if they are not able to understand. Each senior member of WBSEA took the responsibility to train and assist one of the junior trainees so that they concentrate on one field at a time and rigorously help the person to learn effectively and quickly. Such a concerted attention is required to impart the skill as the trainees are totally illiterate.

After the first fortnight of the commencement of the training, the trainees started realizing the complexity of the technology in put, frequent visits of the visitors at the national and international level at the solar house in rural technology park made the trainees nervous and the going was getting tough day by day. A kind of fear was engulfing them as they were new to the atmosphere and each time a visitor met the trainees, they felt that they have a very big responsibility over them and were becoming apprehensive if they will be able to meet the targets & expectiations of the people. With this stress in their mind the trainee Papayamma complained of Chest Pain, while Kumari and Santhoshamma developed fever which never subsided. The subsequent and continual illness worried the WBSEA and the trainee girls were sent for treatment at NIRD health centre. But the trainees were unable to recover and so they were sent for further tests in the corporate hospital. Till then the APTRIPCO had not released the finances and it was difficult to meet the expenditure on treatment for the trainees. However small grants of Rs. 2000 were released by the NIRD-RTP upon constant lobbying for the release of some funds for the maintenance with an assurance to reimburse the amount as the money from A.P.TRIPCO flows into WBSEA account. Thus immediate needs grew day by day in the solar house which needed to be addressed immediately then and there itself and there was no tomorrow. From the small allowances that were made to the WBSEA all the medical expenditure was met.

The trainee, Papayamma suffered from very high temperature which was reported late in the evening and she was sent to the lady Medical Officer in the health Centre at NIRD, but by then the health Centre was closed. However the lady medical officer was to treat the patients out of office hours also for which the institute has made special provisions to attend emergency cases. Realizing the import ance of medical treatment, the trainee was sent for a checkup with a doctor by paying the consultancy and was treated by Medical Officer outside the campus the next day. The tests reported normal and the girls picked up good health in 10 days. We were relieved with the conclusion that this was purely a psychological disorder and the trainees again got into the routine work and started getting on with the training once again.

Every day the trainees started their schedule at the solar house at 9 am and every one was cooperative in carrying out the daily chores at the solar house like cleaning the solar house, filling water etc. One would clean up the solar house, one would fetch water. While the other two would water the plants.

3.17 (A) Financial Constraints in Organizing the Training

Day by day it was getting difficult to meet even the minor demands of the trainees as the finances approved for the project , for training were not released, thus a proposal was put up to the DG/ DDG to provide some amount to meet the immediate needs of the trainees. The DG was very considerate in releasing an amount of Rs. 1 lakh immediately. This amount enabled WBSEA to meet its basic requirement.

After the registration of the society, the DG insisted on designing a logo for solar house. Accordingly the computer analyst of NIRD was called in and provided a brief idea of the proposed logo. Within a week few designs in draft forms were made available by the analyst. Of the designs displayed the BSEs were asked to select an appropriate logo. The BSEs finalized the logo depicting a rural women carrying the solar lantern, the DG was not much happy and the DG, DDG and the Prof. & Head CTPM had a meeting in the evening with the computer designer Mr. Bhatt, this time a logo showing a traditional women with her hands full of bangles holding a solar lantern' attracted the WBSEA and this was finalized as WBSEA logo and a letter head was printed with the selected logo.

3.18 (A) Second Solar Electrification Project for WBSEA

In the month of September, WBSEA received a project from NIRD to electrify the rural building centre consisting of 14 model buildings of different states across the country. WBSEA was to connect all the model houses through the 5 kW power plant. NIRD has also released an amount of Rs. 2, 85,911 to carry out the works related to the project and expenditure towards it. There was no official order but the amount was released and credited to WBSEA account. The solar electrification of all the 14 building models in the Rural Building Centre was completed and made ready for the foundation day in the month of November.

At the solar house the trainers had lot of work in their hand; they were making 12 v l0w solar lanterns and were also training the Paderu trainees. On the other hand the BSEs had to start the electrification of the rural building centre, the work order given by NIRD. The electrification of the Rural Building Centre was an immediate task that the WBSEA women has to take up and they made an inventory of the all the items required to start the project and procure the materials for which money had to be withdrawn from the WBSEA account. There are two Executive members in the WBSEA association. One who is initially an executive member has the signatory powers to sign on a cheque in association with the treasurer of the association as it is a joint account, while the second executive member paired up with the president constituting the second set who would operate in the absence of any one of the members in the first set. But the executive member is given the responsibility to release the funds for the business activities of WBSEA was not showing much interest in the solar house and its activities rather her attitude hampered the progress of the works as she was not releasing the money in time and the works were getting delayed. It was thus getting difficult to take her signatures on the cheques. The Executive member was either not available or she started raising a lot of queries on the amount that is to be withdrawn and started instructing the bank not to release the finances as proposed. None of the executive members do possess any technical knowledge on the solar components but as it is a budget proposed by the WBSEA in association with the technical assistant, release of the finances is imperative. The WBSEA treasurer, being troubled, met the second executive member, who perceived the issue as genuine and passed the cheque, lest the work would suffer and WBSEA will not be able to complete the work in the stipulated time. A cheque for Rs.40,000/ was signed and sent to the bank but the manager refused to give the amount in bulk. Again the cheque was cancelled and instead , a demand draft was drawn to procure the products. The explanation of the bank was that withdrawal of such huge amount will invite malpractices. Additionally as this money belongs to the society a demand draft would put us in a safer stead and enable smooth financial management. According to their request the WBSEA started working in lieu of the guidelines so provided by the bank. As the second executive member was creating much confusion DDG and Registrar asked WBSEA to drop the Executive member from the association.

The training continued and the girls got adjusted. In the month of October the trainee Pravalika and Kumari`s husband visited them and stayed for 2 days and left.

On the eve of NIRD foundation day celebrations on November 9th it was declared by NIRD to celebrate the foundation day for three days, on this occasion the CM Dr.Rajashekhar Reddy visited the rural technology park and it was the most exiting occasion for the 8 BSEs in the solar house. The previous day, the WBSEA president went and bought the marigold flowers and the whole night they knitted the flowers together and decorated the solar house. On the day of the CM`s visit, the 8 BSEs took time to dress themselves in their choicest attire and adorned themselves traditionally with vermilion on their foreheads, and lot of bangles in their hands. For the inauguration 4-5 ministers also arrived who went around in the solar house and Brig Reddy, Consultant, RTP, took the lead to explain them the details of the project. AT 6.30 pm the Chief Minister had arrived at the solar house. The BSEs received the CM and introduced themselves, after a small introduction and briefing on the activities of the BSEs in the solar house by the Project Director, Rural Technology Park. The CM intervened and asked Yellamma one of the BSEs, who was holding a Primary Circuit Board (PCB) in her hand, the names of parts of the components assembled on the PCB, she recited all the names of the components such as transformer, Pot core, etc. the CM was amazed and gave a hearty laugh in appreciation as he did not expect an illiterate women to pronounce such technical terminologies. The chief Minister then asked her about the wages that they were receiving, she told him that they are able to earn an amount of Rs. 2,500/ month from NIRD. Then the CM spoke to another BSE - Papayamma the trainee and asked her how she would test the lights? She showed him the process and he was very happy. At the end of his brief visit the CM wrote his comments in admiration of the work done by the illiterate women and applauded NIRDs efforts in rural development in the visitor's book.

Following the visit, the 8 BSEs were called on to the stage and felicitated by the CM by honoring them with a shawl and a memento. This has been an eventful day for the eight BSES.All the proceedings were aired on the local news channel and the WBSEA women were appreciated and this raised their levels of confidence.

Upon successful completion of 4 months of training the names of the 4 trainee BSEs were also included in the WBSEA association and the new members list is shown in the table in annexure -3.

By the end of November 2005, the first installment of Rs. 2 lakh was released by Tribal Welfare Commissioner. With this release the work related to the assembly of 132 home lighting systems for Thamingula and Poosalapalem villages ensued. The trainee BSE's got adjusted to the new atmosphere and now solar technology did not scare them rather they were well versed with all the details of the technology and they were emerging as confident and empowered women, not only technology wise but also personality wise, as were seen discussing on various issues of the society with the other trainees and could explain the details of the technology and the project to the visitors coming to RTP- NIRD.

The trainee's husbands made a second visit to RTP and left. They arrived without any information and left on their own. After their return to their respective village, they have written letters to their wives. As none of the girls were literate, the letters had to be read out by Papayamma who was semi-literate. However the contents of the letter were little disturbing, where in the crux of the letter showed devious comments made by their husbands in all the letters towards one of the trainees by name Papayamma. In their letters they (Husbands) were vehemently asking them not to interact with Papayamma and threatened them that in its event they (the husbands) would desert them (other trainees). As similar discussions emerged in all the letters, Papayamma started crying since she was virtually reading out all the letters to them. This behavior in the trainees has been the off shoot of the good will and appreciation that Papayamma gained at NIRD. Jealousy overtook the trainees who made comments assassinating the character of the trainee in question. The trainees out of jealously wanted to malign her character and spoke to that effect to their husbands. Sensing the seriousness of the issue, the WBSEA Trainers called back all the husbands of the trainees and discussed out the issue and settled the problem amicably which seemed very silly and immature. Apprehending that such a behavior might engulf the WBSEA trainees, the WBSEA instructed them (husbands) to keep themselves away from visiting their wives until and unless they are asked to come to the rural technology park.

3.19 (A) Initiation of Solar Electrification Project in Tribal Villages

As part of one of the objectives of the project, WBSEA was to conduct training and awareness programmes to the Village Environment and Energy Committee (VEEC). In addition there was a need to mobilize the households to deposit the amount for installation of the systems. After the completion of three months of the training programme the WBSEA team visited the village as the village community assured them of making the payments (registration amount of rs.1000/-) as this was a cropping season and they would market the products and the households will have enough money to pay for the installation of the systems. To meet this objective the WBSEA planned to go for VEEC training on 14th December 2005 and the team comprising of the four Paderu BSES, the young professional, technical Assistant, President WBSEA and the Executive Member left for Paderu on 15.12.05 by road, and reached Chintapalli on 17.12.05. It was a very cold December time and all efforts made to keep one self-warm was futile as it was a hilly area. Our travel with inadequate preparedness to carry warm clothes put us to bare the chill in advertently. WBSEA got the accommodation in the Roads & Buildings (R & B) Guest House to stay put during the VEEL training period. Here the temperature was around 18 degrees centigrade and there was also mist and due in the atmosphere and the viscosity in the atmosphere was increasing against the time and visibility became poor till 11 am. On 18.12.05, early hours, the Guest House keeper informed us that the naxalities visited the place in the early hours and took information about us and, the details of the team (WBSEA) taking accommodation in the guest house. Being perturbed with the information the team became cautious and with all apprehension traveled into the village the next day.

On the first day the WBSEA team visited Thamingula village and all the members of VEEC arrived and WBSEA team discussed with them of their welfare and took stock of the information of the number of households that have paid the installation amount (Registration amount - Rs. 1000). It was seen that out of 125 households that are to be electrified considering both the villages only 80 households had paid the installation amount (both the villages put together). WBSEA was confident that the households will come down and pay the installation amount and waited for another 2 days expecting the beneficiaries to pay the installation amount, but as this was not materializing WBSEA took a decision to meet all the households individually and discuss the problems hindering them in making the payments. Payment of the installation amount was crucial for the project and if the households do not make this payment, sustainability of the programme will be at stake and the basic purpose of the project gets defeated. WBSEA was nervous at this juncture since we were not able to assess the pulse of the households. Therefore the WBSEA team personally visited each and every household in both the villages and enquired the reasons for the delay in the payments. Upon discussions, it was found that the tribals are very much interested in installing the technology in their households and also in their village but the delay as explained by them is due to non-availability of money with them as they still have not marketed their Rabbi crop. Usually they grow and market Rajma seeds in their rabbi crops. As they were still harvesting the crop at the time of our visit, the households could not make the payments. In addition the villagers started expressing doubts on the working of the system, as to whether it will work-in the rainy season? Whether the systems are genuine? By and large each and every one was waiting for the other to make the payment.

These apprehensions were bogging them down and they were shy to discuss the issues with the WBSEA. On the second day, the team visited the Poosalapalem village which also pronounced similar results and not much activity has gone in the collection of the installation amount. On the third day the team met the MDO and the local MLAs and informed them of WBSEAs solar electrification programme. And permission was sought from MDO to install the rural electronic workshop in Thamingula village at the panchayat building. On the fourth day the signatories for handling the bank account were called to the mandal headquarters at Chintapalli and the bank account was opened for both the villages. On the 8th day after a thorough understanding and discussions with the households and the panchayat officials, WBSEA could gauge the gaps and the problems in the village and the immediate emergent needs to be addressed.

Therefore WBSEA organized VEEC training for all the households from both the villages by incorporating the issues to be addressed in strengthening the project objectives. As part of the training some case studies on management of solar electrification programme were presented to the villagers by screening the films through a projector and LCD player hired from Narsipatnam. The next two days, the VEEC training was in full swing and the people were provided technical details on the solar lanterns, solar home lighting systems and its types, the solar pumps and the management of the solar systems by the VEEC. The meeting was attended by small children, men women, young and old. Every day the training lasted till 11 pm. In order to make things clear and audible, a mike set was also hired. The four trainee BSEs also spoke of their experiences and the advantages of the technology in the village .Then the youth and the men interacted with

WBSEA and clarified on all their doubts. The youth of the village came up with three pressing questions:

1. Will the panels work during the rainy seasons?
2. Is there a provision for street lighting system?
3. Will the children be able to read in the light so provided?

All the questions were discussed in detail and answers were provided to the satisfaction of the tribal households by the trainee BSEs and the WBSEA.

At the end of the training, the households in the villages came forward and reaffirmed their consent to install the technology in their households. The panchayat officials including the sarpanch of the village, it's neighboring villages, the MLAs, the government officials like the ANM, veterinary Doctor etc. also spoke and strengthened the confidence of the people. And WBSEA was happy to see the new confidence in the people and their preparedness for the commissioning of the programme which was scheduled in the month of March 2006.

The team returned back to NIRD on 22nd of December 2005 and started preparing for the assembly and installation of the project, which was decided to take off by 2006 March .The solar barefoot engineers, started collecting the quotations for the procurement of the materials. Within a span of 2 months all the systems were assembled by the 8 BSEs.

WBSEA was busy assembling the systems and the message of metamorphosis of the illiterate women into barefoot solar engineers started making news with the media, scientific and bureaucratic community. The BSE's were racing against time to complete the work as the date for installing the systems was fast approaching. It was during that time, much Government and Civil Society Organizations started showing interest in the technology and queries started pouring in WBSEA office and details were sought on solar electrification with different configurations of the panels, batteries and the number of lights that can be provided in each configuration. WBSEA started answering all such demands and it gave much scope to WBSEA to learn about the technology and it became confident on designing the projects .The Director General and the Deputy Director General too started seeking costing details of various configurations of the systems as requested at the higher levels. WBSEA gracefully answered all the queries, but it took away a lot of time and WBSEA had to work for extra time in the nights to complete the pending works.

By the end of February 2005, the training concluded and all the 4 trainees were sent back to their village. The four senior trainees were left in the solar house to complete the remaining loose untied works .WBSEA was left with just 10 days to move into the village for installation, every day a big list of 'things to do' were floated by the team and the works were mounting day by day and as the WBSEA is not able to complete their works, each day they had to carry forward its unachieved targets to the next day, as a huge task was still left to be completed and the manpower was limited. In addition, internally amongst the four BSWs one of the BSE by name Ms.Kalavathy was busy in a family function so only three BSEs were left, of them the BSE by name Zahida is very slow, and ultimately only two of the BSEs took the responsibility to complete the work. On the last day the BSES started packing all the materials. Only two continued to work till late in the evening while third BSE who has a bad habit of taking to drinks, left the workshop as it was time for her to have the drink. The rest of the members stepped out and brought her back to the workshop. The executive members and the young professionals threatened to complain against her behavior to the DG and finally made her work along with the team to complete the works.

Zahida was cajoled, threatened and boycotted for her behavior but she refused to mend her ways. Failing in its efforts to rehabilitate her, WBSEA accepted her as she is. Then all the WBSEA members along with the young professional and the executive member spent the whole night in the solar house and packed the material. By the time the packaging was completed it was 4 am. The next day the transport services were called for to transport the assembled material to the village. The vehicles promptly reported at the solar house to load the material .The loading started at 11 am and it was 6pm by the time the loading was completed, and packed. WBSEA took care to meet the needs of transportation of the products and handed over the details of the transportation of the consignment to the driver of the vehicle lest they be detained on some pretext or the other.

3.20 (A) Installation of the Solar Systems

The total assembled material for the entire project weighed 10 tons. The material had to be adjusted in two different vehicles and they were carefully packed and bound by a rope to the vehicle itself. On 9-3-2006 the ten ton material consisting of assembled and fabricated home lighting systems to be installed in Poosalapalem and Thamingula villages in Paderu ITDA mandal were loaded and transported to the villages from solar house RTP, NIRD. The technical Assistant and the young professional accompanied the vehicles to the villages so as to transfer the systems with care and safety.

The WBSEA team comprising of the four barefoot solar engineers, technical assistant and resource person, young professional and executive member-WBSEA left for Paderu on 11.3.2006. Subsequently WBSEA tried to contact the Project officer, ITDA on their way to Paderu. But, as the PO was on tour the WBSEA team left for Thamingula village. It was 4.00 pm by the time the team reached Annavaram village which is 7 kilometers away from Thamingula village.

From there the road is a ghat road full of serpentine bends and the structure of the road is rocky. Steep edges of the road made it very difficult for the vehicle to climb up the road. Mid-way through, the vehicle stopped and the driver reported that there was no diesel in the vehicle and that he has exhausted the diesel completely.

However the second vehicle which was trailing behind had a reserve of 15 liters of diesel. It was clear that the drivers were playing the trick as the roads were bad, and , in order to extract some money, they seem to refuse to take the vehicles to its destiny. However, as it was our responsibility to transport the products, WBSEA, took the initiative to fill in 15 liters of diesel so that the vehicle can reach the village.

WBSEA had to wait for 1 hour to get the diesel from an unauthorized place from where diesel was procured in loose quantity, however the price was very high and WBSEA had no option but to pay for it and get going. After filling in the diesel the vehicle started moving. But as the drivers were new to the place they did not know how to drive on the hair pin bent roads and they were unable to drive on the ghat roads and refused to go any further and again the vehicles stopped. Seeing the WBSEA members struggling with the drivers of the vehicle, up the village, the villagers requested another vehicle driver who was coming down from Thamingula village to guide the drivers (from Hyderabad) in driving on the ghat road and reach them to their destiny.

Each of the transport vehicle was carrying 5 tons of solar home lighting material and it was too heavy to take the load up the mountain of the village, especially the vehicle which was carrying the batteries was little heavier and it refused to move after a short distance rather the back tyres of the vehicle started slipping as the road became much slippery with the rain and the rocky terrain. WBSEA was filled with anxiety and fear as the vehicles carried very delicate and expensive solar modules. Any breakages would again call for an extra expenditure to procure the new components. All these fears upset WBSEA members. Additionally as it was a naxalite area and as it was getting dusky, fear of naxalite interference started bothering the WBSEA.

WBSEA negotiated with the local driver who came from Thamingula village to shift few of the batteries from our transport vehicles into his vehicle. However, this driver also demanded a lot of money to transport the vehicle. Though the villagers requested him he agreed to transport the products at a cost of Rs. 400/-. Lacking an alternative, WBSEA agreed to pay the amount demanded and transported the solar material to Panchayat building in Thamingula village.

As the vehicle started moving again it got struck into the stones and refused to move further. The villagers again got down from the vehicle and put boulders behind the back tyre so that the transport vehicle will not roll backwards as it starts moving forward up the road with a heavy load on a 60 degrees gradient. All the villagers pushed the lorry from behind and encouraged the driver to accelerate completely. The driver did so and the vehicle started moving slowly. Again after a short distance the vehicle stopped because of big rocks on the road and both the sides of road were steeply cut. Then again the villagers and WBSEA members got down and monitored the vehicle and the driver and the local driver got into the vehicle and drove the vehicle half of the distance. After crossing a short distance the driver from Hyderabad took over and slowly drove the vehicle to Thamingula village. By the time the transport vehicles reached Thamingula village it was 12.00 am. At that part of the night all the villagers came out from their houses and helped us in down loading the material into the panchyat building which was kept ready by the trainee BSEs. WBSEA then made payments to the local driver and the local people who helped in transporting the solar material to the village. WBSEA then left for Chintapally and stayed in the agriculture university rest room situated 30 km from Thamingula village as there was no place to stay in the village. It was 3.00 am by the time WBSEA settled down to rest.

Devoid of a suitable place to take shelter in the village WBSEA team commuted from Thamingula to Chintapally every day. The local people came to the rescue of WBSEA who offered a small room of the size 6 x 9 for accommodating the WBSEA. All the four BSEs and the executive member slept in this room for the rest of five days and the technical assistant and the resource person and the young professional rested in the panchayat house making a small space for themselves in the solar house amongst the solar material. The trainee BSEs organized the food arrangements for WBSEA team. The food grains were procured by the team for cooking in Annavaram village. The very next day that is on 13-3-06 the WBSEA team reached Thamingula village by 11.00 am. However, they were surprised to hear that every Monday was a market day for the village which they called it as 'Santha' in local language and this was organised at Annavaram village. All the villagers left to procure food grains and essential requirements for the family for the whole week. WBSEA had to wait till 3.00 pm to start the work.

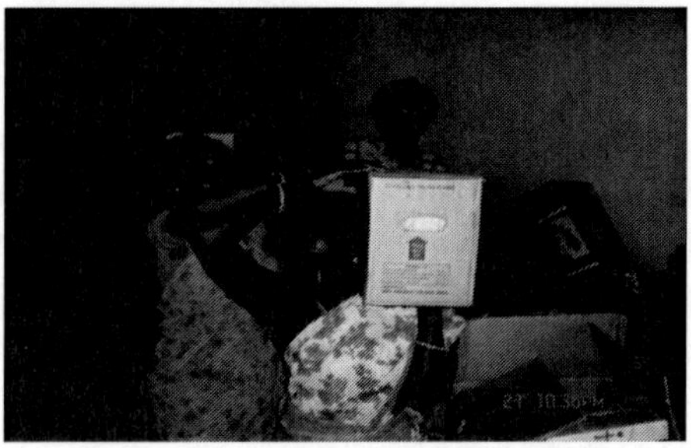

The households that have already paid the installation amount of Rs. 1,000/- to VEEC were given the solar material by registering their names and comparing them with the names provided by the village energy and environment committees. On the first day came forward and 10 members procured the solar material. All the heavy material like the poles, batteries, panel, the sleeves for module mounting were given to the villagers by recording (is a register), each and every material provided to them. Later the WBSEA BSEs went and installed the home lighting structures in the households.

It was surprising to see that our four BSEs and the trained four BSEs put together could install only two home lighting systems on 13.3.06. This situation gave much ambiguity in completing the target within the targeted period. Amongst the senior BSEs from Hyderabad

Chennamma the President and Yellamma the Treasurer were very proficient and skilled and showed complete responsibility in taking up the work and just took an hour to fix a single home lighting system. However the rest of the other BSEs Zaheda and Kalavathi and the trainees Santoshi and Pravallika stood as on lookers and did no job except watching the work done by Chennama and Yellamma. Observing their behavior it was felt that these BSEs lack confidence in fixing the systems.

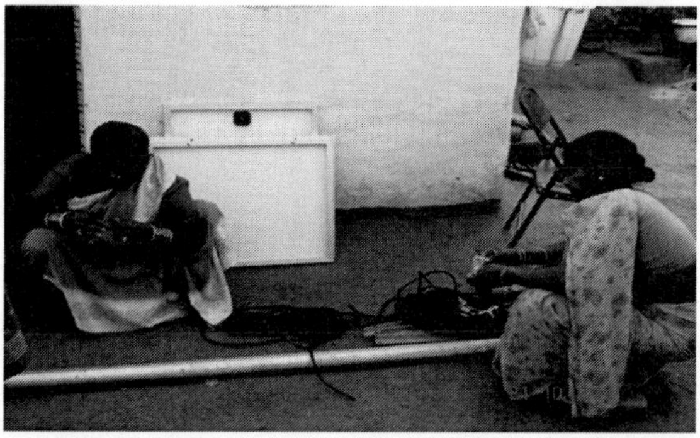

The next day it was decided to pair one senior with a junior. The teams working with Yellamma and Chennamma could fix up home lighting systems in four houses the next day.

Of the other two groups Kalavathi did home lighting for two houses in a day and Zaheeda could not fix up the systems in even a single house, she spent the whole day on a single house and that also

she was not able to complete it successfully and could not illuminate the lights and at the end of the day then Chennamma, intervened to correct the system and the lights glowed.

Following table gives the details of solar electrification in the selected village;

Sno	Item	No of HHs	Observations/progress
1.	Total number of households to be electrified in the two villages	132	• -As per the approved project • -There are 128 households living in the villages as on 13.3.05.
2.	No of houses to be electrified in Thamingula village	88	• Only 66 Houses have been electrified in Thamingula village as on 18.3.06. •-The remaining 22 houses could not arrange for Rs.1000/towards installation charges as agreed upon. •-These 22 households have been promised a loan of Rs.1000 through SHG, by the PO -ITDA
3.	No of houses proposed to be electrified in Pusalepalem village	37	• Only 25 houses have been electrified in Pusalepalem village. The remaining 12 households also have been promised a loan of Rs.1000 through SHG, by the PO -ITDA.

The electrification and its advantages spread in the surrounding villages and also in the study village through word of mouth and observing the lighting systems in the neighbors houses , a lot of people turned up and collected the material by paying the installation amount (Rs.1,000/-). People became very enthusiastic to illuminate their houses the very day that they took the solar home lighting systems to their houses. But, our BSEs were not fast enough in meeting the demand. Then the BSEs were motivated of provide an incentive by way of adding an incentive to the salary payments based on the number of houses that each team will electrify in a day.

Subsequently in a single day, Chennamma completed installing solar home lighting systems in six houses and Yellamma did it in seven houses while Kalavathi did two - three houses and Zaheeda remained by home lighting a single house with great difficulty. This made the WBSEA President to probe into the problems of performance in the non-performing BSEs. And this called for a serious decision by the WBSEA in the future to accept only those persons for training that are skilled and not to take any chances. The WBSEA took time to understand the fact that the selection of the trainee is very important. Thus Kalavathi and Zaheeda were asked to step up their skills and performance. Because of the lack of confidence in the skill they also lacked the commitment as was shown by the President Chennamma and Treasurer Yellamma.

After the completion of the installation of home lighting systems the rural electronic workshop was installed at Thamingula village by the BSEs. It is complete in totality and has been functional since 16.3.06.

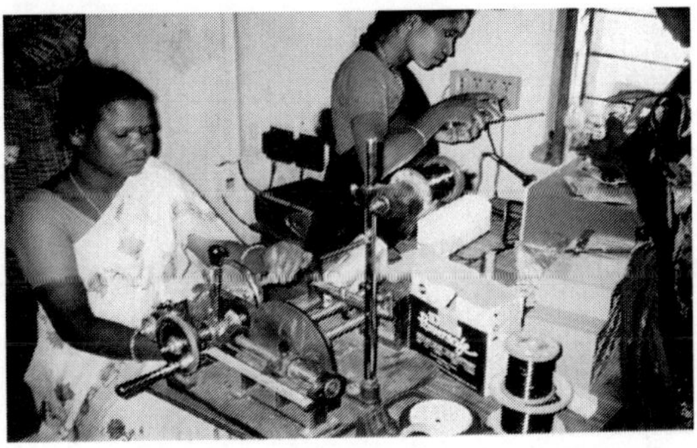

3.21 (A) Peoples Struggle for Electrification in their Households

As home lighting was completed some of the households came and sought permission to make payments (the installation amount) in installments as they could pay only a sum of Rs 500-600 only at the moment, while some others expecting their money shortly also wanted to take the systems and install them on an installment basis. As a clear stand could not be taken on this issue a panchayat meeting was held on 15.3.06, to seek a consensus on the issues. In the panchayat meeting it was decided that solar home lighting should be provided only to those individuals who would pay the complete amount of Rs.1000.

One of the women belonging to the Primitive Tribal Groups (PTG) was serious in installing the lighting system in her house. Her name is Korrapalu. Korrapalu's husband was away in the town for an employment, she was expecting that her husband would return back with enough of money to make payment for the installation of the system. But to her dismay her husband returned empty handed and she could not make the payment for the system. She sat disappointedly and was hovering around the panchayat building where we down loaded the solar systems. Perceiving her interest in acquiring the system for her household the village elders quickly resumed a meeting as she is a very sincere person and decided to fulfill her desire by asking her to take care of the sarpanch's cattle for the whole year wherein they will be paying only a part of her wages at the end of the week, however the sarpanch has agreed to provide her meals for once in a day. Korrapalu agreed to the decision taken and happily took her systems home and the sarpanch came forward and paid the amount to VEEC.

In most of the tribal households only the women go to work while the men are intoxicated throughout the day and stay at home. Such families found it very difficult to pay the installation amount for the systems. In one such instance Lalitha a tribal women who wanted solar light in her house gave a witness of her turmoil who thought that it is not impossible for her to meet the demand, so she fasted and prayed the whole night for the money and the next day she traveled 60km and went to her mother's place and discussed her problem with her father and her father gave her the installation amount and she quickly travelled back the same day and picked up her home lighting systems and installed them.

In another instance as the WBSEA, BSE'S were busy installing the systems, a small child watched the whole installation process at his neighbors place and was excited to see the lights turn on in a dark room. Immediately he went into his house and asked his father to get the solar system fixed at their residence also but the father replied that they do not have so much money to afford for the lighting, when his father did not heed to his polite request, the child started crying and rolling on the mud and attracted the attention of everyone. Unable bears the sons protest the father immediately made payments for the installation of solar system in his house. Few other households quickly went to the market, sold their produce and arranged for the finances.WBSEA could observe the spread effect in the village with one family buying the system and the others emulating behind them.

In one another instance as BSE's were installing the system in a household the neighbor came fully drunk and seeing his neighbor getting the systems installed he started questioning his courage to get the system installed before he himself has done it, with all serious protest throughout the night, until he exhausted and the next day he came forward, paid the amdint and took his system. Many women were auspicious who decided on the time and date of installation and the person that would first operate the system.

On the eastern part of the village there are nearly 30 households belonging to the Primitive Tribal groups who have not come forward to register themselves for the solar home lighting systems for their houses. WBSEA visited the households and held a meeting in their premises. The PTGs are very poor people and they genuinely belong to the below the poverty line status. Their women and children are barely clothed and they live in semipukka houses and huts. They spend more than 6 months in the forest areas and depend on forest produce for their livelihood.

The PTG's insisted that the solar home lighting be given free of cost to them as they are and will not be in a position to pay the installation charges of Rs1000 immediately and Rs.100 per month. The PTGs started bargaining that they will not pay the installation charges but will only pay the user charges that to only Rs.50 and not

Rs.100. However as their houses are situated closer by WBSEA decided to club two houses together and provide a single bulb for each house, under such circumstances the households can pay only Rs.50 as user charges.

The next day, on 16.3.06 the PO - ITDA visited Thamingula village when panchayat meeting was going on. Upon appraisal of the situation the PO has agreed to release a loan of Rs.1000 to all SHG members that cannot afford to pay the installation charges immediately. The amount was proposed to be recovered in 20 installments. Then consent of all the households that are willing to avail the loan and subsequently adhere to its recovery systems was taken and the list was handed over to the APM for the release of the loan amount.

Upon reaching Hyderabad all these issues were discussed with the Commissioner for Tribal Welfare who agreed to release an amount Rs 37,000 for the PTG 's and facilitate in solar home lighting for their households.

Being satisfied by work done by WBSEA, Commissioner has asked WBSEA to put up the file for release of finances in Phase -II of the project for establishing 1 KW Power plant for solar street lighting and community TV in Thamingula and a community TV in Poosalepalem village (details of financial allocation received from ITDA paderu by WBSEA enclosed in annexure 5).

3.22 (A) Other Projects of WBSEA

Solar electrification project in paderu mandal gave enough leverage to WBSEA to take up various other solar electrification projects. On a turnkey basis WBSEA had been receiving projects for solar electrification and in the meantime WBSEA also started producing new projects to meet the demand of the market.

3.23 (A) Solar Street lighting Project at NIRD

Following the completion of these projects, WBSEA received another project from NIRD for solar street lighting systems in the NIRD garden at the jubilee park, before the administrative building. The then DDG Shri. Chinmay Basu was very keen in helping WBSEA. He asked the WBSEA to provide quotation for a single street light. Straight away WBSEA gave MNES specifications and the costing details of a solar street light which costed RS.28, 000. DDG was not happy with the proposal and called the WBSEA team for discussions again and gave instructions to reduce the cost by minimizing the technology to suit the garden .WBSEA had to apply its mind and came out with a proposal to mount a 10 watt module on the pole . And the cost was brought down to less than 50% and the proposal worked and WBSEA got the project for installation of 7 solar street lights in the garden. This way WBSEA was also involved in the research and development aspects of the technology .WBSEA was fast in calling quotations and it installed the systems. The lamps successfully lit for 15-20 days and suddenly stopped .WBSEA was seriously criticized for its unanticipated faults. It took some serious brain storming to understand the inert problem. WBSEA recognized that it has used a circuit from another company. For effectiveness and efficiency the circuit and charge control should be separated but as they did not exist in independence and the problem persisted and compounded, affecting other parts in the circuit also. WBSEA thus realized its mistake and changed the systems and arranged for the independent existence of the crucial components. This rectified and regulated the lighting system and saved WBSEA from further humiliation.

3.24 (A) Training the NGOs

The Deputy Director General arranged for a meet on 5th of May 2006 for all the NGOs working with the objective to promote renewable energy interventions in their project areas. The meet was designed to bring in collaboration between WBSEA and the NGOs and market the products into the rural areas. Very famous NGOs and manufacturers of solar technologies attended the meet which was chaired by shri.T.L.Shanker, the former Director General of ASCI. The workshop was organized to discuss means of transferring solar energy technologies into the rural areas through the NGO`s. The purpose was to make the NGO`s partner with WBSEA in transferring the solar energy technologies to the villages.

After detailed discussions it was felt that a TOT (Training of trainees) may be provided to educated rural youth, who would latter provide servicing of the systems. Some NGOs also expressed their interest in establishing rural electronic workshops in their villages with the assistance of WBSEA, for which WBSEA assured them support in establishing REW, and training in operation and maintenance also.

Most of the NGOs were interested in training in solar technologies but expressed their inability to raise funds for the training. This issue was discussed within NIRD and it has been agreed that we may provide training to the interested NGOs free of cost under a package. Accordingly a letter was sent to nine NGOs, asking them to select some rural youth who have some basic qualifications from ITI institutions for training at WBSEA - NIRD. Youth from 5 NGOs responded to our proposal and came forward to take the training.

3.0 (B) Solar Electrification Project Initiated at Tingnaput Village

Learning from the experiences of WBSEA -NIRD and in a bid to empower the rural women in solar energy project as a part of the livelihood programme, the Orissa Tribal Empowerment and Livelihoods Programme has requested for the services of WBSEA in solar electrifying 30 household in the state of Orissa, Tingnaput Village. Accordingly they selected four illiterate women hailing from tingnapaut village in Koraput district from the state of Orissa. These women were selected and trained in solar technologies at the solar powerhouse -National Institute of Rural Development, Rajendranager Hyderabad.

3.1 (B) Initiation of Solar Electrification Project in Tingnaput

The sole idea of empowering the women in the village was in response to the previous experiences of investments in developmental activities which have established unsustainability of the programmes due to lack of servicing facilities upon its breakdown. Thus in order to make the programme successful and sustainable, the selected Tribal girls have been put to training, from 16th of August 2006 and are taught to fabricate solar lanterns, assemble home lighting systems and street lights, so that all the breakdowns and failures of the systems in the future can be addressed to by the trained Tribal women themselves, who after the completion of the training are called as Barefoot Solar Engineers (BSEs).

The four women were Talso, Phulka Wadeka, Bundei and Meenakshi. The four trainees returned from six and a half months training and they have mobilized their village for the installation of 32 solar home lighting systems in the village with an initial contribution of Rs.500 as installation charges.

After scrutinizing the proposal for solar electrification in the village, that was proposed by WBSEA -NIRD the commissioner for Scheduled Castes and Scheduled Tribes has released an amount of Rs.6,99,633 lakhs, towards training four Barefoot Solar engineers, home lighting systems for 32 households, one Rural Electronic workshop and Village Energy and Environment Committee training. Each home lighting system costed Rs.9, 010 which provided a 20-Watt solar Panel, 1 CFL tube light of 9 Watts, and an LED concentration of 1milli amperes and 5 streetlights.

In addition OTELP has also provided an amount of Rs.44000/ for 200 days @ Rs.55/day as wages for the women since they will not be able to earn any livelihood during that time. VEEC (Village Energy and Environment Committee) was formed on 6th March 2007. They have opened a VEEC account in the State Bank -Narayanapatna.The VEEC is named as WADEKA RAJU JANI. Today the VEEC handles their own accounts and records. The Village Energy and Environment committees (VEEC), comprises of the Self-help group Leaders and the panchayat members who monitor the installation and maintenance of the systems in the future. As on date an amount of Rs. 5600/- is deposited in their account.A unique feature of this project is that the villagers have agreed to contribute an amount of Rs.700 / household at the time of installation of the home lighting system, and Rs.50 every month /house hold after the installation of the home lighting systems. Based on the nomination of the villagers two individuals from each village are given the signatory powers to operate the accounts. The amount so collected and deposited in the bank, from the villagers is utilized for purchasing new batteries after 5 years as the batteries weaken, and payment of salaries for the four trained Tribal Barefoot Solar Engineers. The BSEs arrived in the state of Andhra Pradesh on the 15 th of July 2006. The training for the four selected trainees started on the 16th of July 2006. The schedule of the BSEs starts every day in the morning at 8:30 am with their Breakfast and by 9 am the training starts in the Rural Electronic Work shop at NIRD. Since the trainees have to know some basic arithmetic and language they are taught Numbers and English for the first 3 hours every day. Following which they concentrated on the solar technologies. The BSEs responded very well to the capacity development programme on the skill and soon were able to read and write numbers from 1 to 10.They also started pronouncing and writing alphabets up to' G'.

As part of the training, the trainees are taught to make transformer and recognize the color codes. Amongst the four BSEs, the BSEs Talso and phulka have been very fast learners. However as learning has been a new activity for the BSEs, initially they were not confident on whatever they were learning, but the seniors took all care to mobilize the BSEs in their skill leanings. It almost took one month for the BSEs to become confident in the skill

The BSEs were grappled with climate change impacts and also the fear of achieving the target and this made them fall sick often. WBSEA took all care to nurse them back to good health by providing them the required medical aid at the Health Center. Within six months the training for the 4 Women barefoot solar Engineers from Tingnaput village, Koraput district was completed and the NIRD -WBSEA proposed an agenda comprising of the following events in the village.

1. Survey of 30 households in Tingnaput village
2. Formation of village energy and Environment committee
3. Selection of candidates for holding joint account in the nearest bank
4. Screening of solar technology based films for the village.
5. Finalizing the points for installation of home lighting and street lights in the village &
6. Selection of site for Rural Electronic Workshop

3.2 (B) Completion of the Training Programme

The scheduled 5 months training period for the 4 Tribal ladies from Koraput district was concluded on the 15th of December 2006. But as the ladies were very slow, and especially the elderly ladies (Phulka & Talso) had eye sight problems, additionally the 15 day vacation in between, all this has slowed down the process of training.

After a discussion with the Technical Assistant and Barefoot Solar Engineers of Women Barefoot Solar Engineers Association at Rural Technology Park, RTP-NIRD, it was decided to extend the training for another one and a half months since the training was not completed and the assembly of remained the proposed solar systems that were to be transferred to Tingnaput village was also pending.

Thus with the extension of the training period the BSEs could complete all the pending works by the 30th of January 2007.

The OTLP was also prompt in releasing the budgeted amount for the procurement of the systems. OTELP also released an additional amount of Rs.18, 000/ towards expenditure for the maintenance of the trainee BSE

3.3 (B) Transfer of Solar Energy Technology in the Tingnaput Village

The WBSEA team left for Tingnaput village, Koraput district, the state of Orissa on 5th March, 2007. Koraput district is 210 KM from Visakhapatnam District of which 82 km constitute the ghat road. Sunki is the last village between the state of Andhra Pradesh and Orissa. The village comes under Podapadhar panchayat.

All through the way evidently the state has very prominently made it a point to document the rural development works on the walls which were completed by the government in NREGS, FFW and SGRY etc.

The Tingnaput village is another 80 km from the Koraput District. The village is situated at the bottom of the hills and the road leading to the village is ghat road with hairpin bends on the road. The scenery was beautiful with high mountains rising on both the sides of the road. However the mountains were devoid of vegetation. After moving a few kilometers, smoke was seen emanating from the mountains. This indicated the traditional podu system of burning the fields soon after agriculture cropping is completed on the mountains. Despite prohibition on these acts of the tribals they continue to adhere to these practices. The forest officer has the responsibility to inspect and curtail the occurrence of all such instances.

Kolabe and Machkund are the two sources of water which emanate from Machkund hills. These water bodies help in irrigation of the agriculture fields in the village. By and large there is no water problem in the village.

The OTELP was launched in Bhubaneswar on Oct 2, 2004 with a prime intention to implement the developmental programmes in the village. For the effectiveness of the Tribal live lihood promotion programmes in the village 4 prominent NGOS were selected to hand hold and give the required support in the sustenance of the programmes. The NGOs signed an MOU with OTEP to implement the programmes. The implementation and monitoring of the developmental programmes in the village are executed by a very reputed NGO called as Vikas. The NGO (Visas) is selected as the representative at the state level from 35 NGOs who submitted their details for the said selection as a representative NGO. The government has shortlisted only 4 NGOs. Of the 4 four VIKAS is one such NGO. The selection of the representative was made very objectively by the participation of a third party in the selection process. The Exim Bank participated in the selection of the NGOs and sent the list of selected NGOs to the government.

The four selected NGOs at Laxmipur are;

1. Ms. Swaminathan foundation.
2. CYS Development society which also worked in Laxmipur village
3. VIKAS worked in Narayanpatna and
4. RAAS worked in Bandhugaom

All these NGOs are considered as the programme implementing agencies while the Swaminathan foundation worked as resource NGO since the year 2004 in the Tingnaput village. The NGO has been very active in providing various trainings to the villagers especially on watershed, agriculture on vegetable cultivation, ragi, paddy cultivation in the Rabi and Kharif seasons since the year 2006. Each training lasted for 2-3 days.

Based on the qualifications and rules set by the government in MORD the NGO engage qualified people like the soil scientist, agriculturist, experts in forestry, microfinance, livestock, engineering etc who monitor the activities in the village based on their expertise. Each ITDA has specialists in these 6-areas who work in unison with the NGO specialists.

The village has formed its Watershed Committees called / Village Development Association which is named Attama. This association is formed by three villages Tingnaput which is a hamlet, Samana it is also a hamlet and Podapadare which is a revenue village. The VD Committee (Village Development Committee) has 19 numbers. It is responsible for entire developmental activities in these villages. Attamma is a perennial source of water body being connected from mountain hills which flow into the village, with which the Watershed programmes are developed.

As a part of the livelihood programme all women have become the members of the SHG since March 2006.

Few educated youth have joined the NGO to work in watershed committees who have been a great source of all help in mobilizing the community.

There are 3 SHGs in the village by the names

1. Sitama peni which has 10 members. They collect an amount of Rs. 25/- per month from each person and deposits it in the bank. As on date the SHG has Rs. 61,000 in its account which started in 2002.
2. Laxmi SHG has 20 members which started on March, 2005 which also collects Rs. 25/- per month
3. Turkiyama SHG started on March 10, 2006 and it has an amount of Rs. 3,100 in its account.

The village is purely a Scheduled Tribe village. Apart from female SHGs in the village there is a single male SHG comprising of the youth of the village. All the SHG members in the village have accessed the bank loan. Sitama SHG took four loans till date. Recently it has procured a loan amount of Rs 2, 50,000 and spent Rs. 10-15,000 on individual business. This business is based on the forest produce which is also called as the Seasonal business. Here the SHGs try to buy the raw materials for making the brooms. The raw material is procured from the forest and the Brooms are produced and sold in the weekly market which falls on Sunday, at Rs. 10 - 12 per broom. Each member is able to produce 40-50 brooms/day and sell 500 - 600 pieces of brooms in a season. Each SHG buys 60 bundles of the raw material bunch which costs Rs.50 which when subjected to its processing will be able to produce 7 brooms from each bundle and sold at Rs.12 each and the ultimate profit is construed to be Rs.34 on each bundle of the raw material. On 60 bundles of brooms it gives 420 brooms per SHG. And when the lot is sold @ Rs.12 per broom the net profit is Rs.5040. And 30 brooms are sold / week and the profit per week is Rs. 1,440. Usually this business starts in the month of November and December by which time they cut and bring the raw material from forest and by April they sell the product after completing all the process of drying, cleaning, mending and tying the brooms with ropes. Usually the SHG loan is spent for consumption. While some SHGs spent the loan amount to procure Lentils, like Kandulu and other variety of lentils which after cleaning them sold them at a higher rate in the weekly market. Apart from these economic activities every member has an access to the forest produce like the tamarind, amla, soap nuts, gum Karaya and other products and collection and sale of these products is an alternative livelihood for the tribals.

One of the SHG member's son has set up a small economic unit in the village which is a general store for the entire village, he is able to bring and sell shampoo, soaps, powders, fancy items for the girls and eatables. Upon an investment of Rs. 5,000/, since the past 6-7 months he is able to earn a Profit of Rs. 2,300.

Despite the harsh toil all through the year, the tribals have a weakness of spending all their earnings in a very famous festival called the Chait parab, in this festival they spend all their money to the extent that they literally become broke. All the Tribals in the village belong to a sub tribe called Khonds. They belong to BPL category, earning less than Rs. 2,500/- year.

3.4 (B) Electrification in the Village

The village depends for its electrification on the Kerosene as a basis for energy. Nearly 1-2 lamps are brought into utilization each day. Two liters of kerosene is consumed each week which is procured @ Rs. 12/- lit and the total budget for the month is Rs.96 towards expenditure on kerosene. Geographically the village is situated at the corner most part of the area, beyond which there are no settlements. At a distance of 1 km from the village almost all the households are electrified along with street lighting in the village. The

Tingnaput village also has been provided with electric wires and poles and a transformer in the village 8 years back, but the actual electricity never flowed through the wires till date and the systems were lying idle. Now these wires were stolen and nearly 500 mts of the wire is lost.

3.5 (B) Housing Infrastructure in the Village

The houses are artificially painted by mixing locally available coloured mud in the village. The Houses are kutcha with tiled roofs and wooden skeletons. The tiles have been provided by the OTELP. The roofs are very low and one has to completely bend down to go into the house. The verandah before the household units are open with large grill windows where the Tribals store their crops. Each household has its own grain bank in the corner of the house. Each house hold unit is a single roomed unit with a verandah in the front and another in the backyard. The middle room is very dark and usually the family sleeps in this place. The OTELP has provided for a complete cementing of the front yard and the tribals keep the place very neat and tidy by applying the cow dung every day in the morning. The houses do not have any locks and the houses are left open and the tribals go away indefinitely to stay in the forests without locking their houses.

3.6 (B) Livelihoods and Funds Expended in the Village

The village has been selected as the model village and funds from various organizations like DFID - 44%, IFAD - 22%, WFP -13%, Govt-10%, Agency (SHG, labor, VDC) have been spent for village empowerment and livelihoods development.

Out of 30 houses in the village none of the households possess any lands while they cultivate on the mountains and depend on rains for irrigation purposes. They usually grow ragi, Mandya, Gatia and 1 Quintal is sold at Rs. 750/- (7.50/Kg). Paddy is not grown in the village since it is in the upland. The Tribals keep a certain amount of the produce for their consumption and market the rest. The following table provides a picture of the produce and the cost at which the product is marketed in the village.

Agriculture Products Produced and Marketed in the Village

Product	Average produce per HH	Rs./ Kg
Tamarind	1 quintal	Rs. 5
Gatia	2-10 quintals	Rs.7-50 Kg
Niger	50 1 quintal	Rs. 20/- kg
Jowar	50 1 quintal	Rs. 7-13
Suwa	5-10 quintal	Rs.7.20 kg
Kandulu	4-5	Rs. 20

The Tribal's usually keep 1-2 quintals of their produce for use and sell the rest while marketing the produce. The product is sold to the money lender/the middlemen. However the Tribal households are not involved in land brokerage since they do not possess any land with them.

3.7 (B) Employment Situation in the Village

For nearly 12-15 days in a month the Tribals in the village get work and they receive wages up to Rs. 50 per day. Apart from it they also go for alternative livelihoods based on forest or NTFP. They are able to earn a Wage rate between Rs. 10 -30/ day. They are also provided food and drink in the place of work. Every household has alternative livelihood systems which are based on livestock. Nearly 4-5 households have cows which are able to provide 8 litres to 15 litres of milk per day. While all the households rear Desi chicks and each Chicken is sold at Rs. 100.

3.8 (B) Social Aspects in the Village

Health wise, the village is average where in men, women and children look healthy. They are able to provide for two square meals for every family member. They consume their food in the early hours of the day between 9-10 am and again have their dinner by 6 pm. There is no child and mother mortality in the village, but while in the rainy season the village gets fregventyl attacked by Malaria and other communicable diseases. In the survey, it was found that only one person suffers from TB and the family is not aware of the health schemes promoted by the government to address the problem of TB in the village. Primary Health Centre is situated in Nagapatana which is 5 km from the village. The ICDS is located at Podapodhar village while there is no Anganwadi activity in the village .The village has a school which is run on the verandah of a house, but recently OTELP has built a new school for the children of the village. The school has 40 children, after school hours the male children go to graze their animals. Girl children do not go to school beyond class 5. Girls do not go to school at all, After maturity.

Marriageable age in the village is 14 years. Boys of this community marry by lifting the girls.

Women smoke cigars by burning it in reverse end when enquired it is found that it gives a good feeling and an effect of the tobacco. Liquor is procured by the villagers from a nearby village called Podapadher, but it is seen that only old men and women consume liquor in the village.

Granary is stored by each household in the first room of the house itself while the adjoining room is used for sleeping, it was interesting to note that the livestock was kept in the last room.

There are 4 persons Educated in the village. Two studied up to class eleven and two of them studied upto class twelve while the other two are educated up to class eight and graduation respectively.

The OTELP has also provided for bore well in the village which is dug up to 200ft, providing for drinking water for the village.

Money lender activity is strong in the village. Every household procures loan from the money lender at the rate of 6% interest.

The village is monitored by an NGOs called VDCT and VIKAS which monitor all the developmental projects in the village. At the time when solar electrification programme was getting launched in the village, watershed programme (sponsored by IFAD) activities were in full swing and VIKAS was monitoring the Watershed activities once in every 15 days.

3.9 (B) VEEC Meeting in the Village

The village Development Committee (VDC) plays a very important role in the village. As already discussed a village development committee is formed by incorporating members from 3 villages. But the villages are all scattered and are distantly placed. For the VEEC meeting the VDC emphasized that representatives from all the villages should participate in the VEEC discussions to avoid all controversies in the future, as is the case with the VDC while taking up any developmental activity. But since the villages were quiet far away and located on the mountain, the village decided not to incorporate any members for VEEC from other villages.

After completing the discussions, the village was ready to make payment towards the installation of the solar systems in the village but felt that Rs. 750/- as installation amount is to too high. Thus the team calculated and gave a new amount as Rs. 700/- per HH to which all the households agreed and affirmed to abide by all the rules and regulations of the committee.

The project director Mr. Deepak Mohanthy visited the village on the second day of the VEEC training and discussed the solar electrification project in the village. The households expressed their payment problems to the Project Director who assured the households that the project will be very helpful for the households.

Educated youth in the village do book keeping where in the members from VLSC, VDC, PRI, NGO and SHG work together .The women are very much empowered in the village as the society is traditionally Matriarchal and the intervention of the SHGs in the developmental activities is evident.

After the formation of the VEEC the committee requested the following Issues to be implemented on a time scale basis:

1. Monitoring and Evaluation every 6 months
2. Early construction and installation of Rural Electronic Workshop Building in Tingnaput village.

An Exposure visit of the village comprising of the SHG leaders was proposed by WBSEA to Paderu project area to provide a motivation to the members of the village. Accordingly all the 30 members, one each from each family attended the exposure visit and the ITDA Paderu was informed of their visit to the village beforehand .The OTELP -Orissa had provisioned to engage 3 commander Jeeps to carry the team, however one of the Jeeps failed on its way as it was climbing the hill leading to Thamingula village. This delayed the whole visit and the scheduled visit was delayed for 4 hours and by the time the team reached the site, it was 4 pm.

Since both are tribal communities the people of the communities got engaged in a dialogue and discovered that the families are related and soon invited each other to their villages and festivals. This interaction encouraged the officials in taking the project forward. By the time the communities finished their discussions it was 6 pm and the households switched on the lights in their homes.

The khonds visiting the village were very happy and visited the village by interacting in much detail and understood all the other technical details of the systems and their working conditions. The visit culminated by 8pm and the team was accommodated at PMRC at Paderu for the night and returned satisfied the next day.

3.10 (B) Interventions of BSEs in Motivating the Community for Solar Electrification

One of the participants Ms. Nisika was a Sarpanch of the village and she took a lot of care in educating the people of the village and motivated them to participaten is the installation of the technology and sought the cooperation of the local community in accepting the terms and conditions of transfer of solar technology.

The village has an Anganwadi centre, and a primary school and midday meals scheme is well implemented in the village. There are 30 children in the village who attend the Anganwadi center and also the primary school in the village.

The Project Director Mr. Deepak (IFS) was very much encouraged by the interventions of NIRD and made it a point to visit the village the next day. In the discussions with the community and the officials it was decided to build a Renewable Energy Workshop adjacent to the school building and install the fans and lights by connecting them to the REW; this arrangement was worked out by the BSEs. The Project Director also agreed to provide three street lights in the village on a pilot basis.

A discussion was held between the BSEs and the members of the village. The men, women and children attended the meeting and discussed at length on all aspects of electrification including its installation, usage of the battery, life of the systems, contributions for the sustainability of the systems etc. Significant of the tribal culture, the women carried and smoked 'bidies' all throughout the day. They smoked for few minutes and put off the beedi and tucked them in their ear lobes (both men and women) and smoked again as and when they desired. Men were always intoxicated and did not understand any part of the discussion but the women were very active and participated in the discussions very carefully and took decisions.

During the discussions some of the people in the village were of the view that Rs. 50 a month, a contribution towards the user charges is high though it has a prudent intention of sustaining the technology. The tribals were later explained by the officials from OTELP in detail on the advantages of the system and the costs incurred usage of on the kerosene. And once the solar systems are installed the kerosene charges are saved and these can be diverted for the payment of user charges which is equivalent or is even less than their expenditure on kerosene, the discussions fructified and all the villagers agreed to participate by paying a contribution of Rs.50 per month as user charges.

After explaining the concept on solar electrification in the village and the various costs that the households have to meet, the village energy and environment committee was formed comprising of the women SHG members and youth in the village. The village is monitored for all developmental activities by an NGO, which has volunteered to initiate the VEEC activities and the collection of the installation amount and its deposit in the bank. On 8th the Director of OTELP had visited Tingnaput village during which time WBSEA was camping in the village for VEEC training. The PD visited the village and discussed the solar electrification proposed, at length for 3 hours with the villagers who assured him to take the programme forward and cleared all their doubts. In the evening, films on solar electrification were shown to all the villagers and explanation on the solar electrification system was provided to them

Based on the discussions held on 8th of March 2007 with the Project Director OTELP the households were sent for an exposure visit to Paderu on 13th. Earlier a small team of the selected trainees were sent on an exposure visit (on July 20th so as to conceptualize the programme process) to Thamingula and poosalapalem villages of Paderu mandal.

On the next day the Project Director planned for an Exposure visit of the village representatives to the technology transferred villages, thamingula and poosalapalem, and the team visited the villages on 13th March and an exposure was given to the members.

The households were very much satisfied with the functioning of the project which motivated them to mobilize their finances towards the installation costs of the solar systems of the households immediately.

Initial discussions on transfer of solar technology and electrification in the village were held with the OTELP officials along with the programme officer at Koraput on 7th March.

After the exposure visit, through people's participation the OTELP officials selected 4 ladies from the village and sent them to the project village foe an exposure visit, they are;

1. Phulka Wadeka 40 years; she is uneducated and has a daughter; she works in the fields and is busy for 4 months (2 seasons) in a year. The village is usually cropping the lentils (Kandhulu) of all varieties. Working in her own fields she goes for labour work for 15 days in a month and she earns Rs. 2,000 / year (Rs. 30/days). In addition to the agriculture work she uses her alternative skills and makes Jhadu by picking up the raw materials from the forest and earns Rs. 300 per season.
2. Mehakehi Diwase studied up to 10th class. Father is a private practitioner (quack) and mother is an Anganwadi teacher.

3. Talso is from a scheduled caste background and works in the Dongori (agriculture), she has 2 children, and she studied up to 3rd class
4. Bundi is uneducated and has no children

Subsequently the trainees were sent to NIRD and the training was initiated by WBSEA and it concluded on 5th of March and the primary assembly of the technology was also completed by then.

The four trainees have fabricated 32 home lighting systems along with four solar engineers of WBSEA, RTP, and NIRD.

On 21st OTELP communicated to WBSEA that the systems should be installed by 24th of March so as to enable the Honorable President of IFAD to inaugurate the solar electrification project in the village.

On 21st of March the systems were loaded and transported to the Tingnaput Village. It literally took 4 days for the consignment to reach the village. But the BSEs worked the whole night and completed the installation of solar systems in 30 households and 5 street lights in the village.

It was encouraging to note that all the people in the village were also awake and participated in the installation of the solar energy technology in the village.

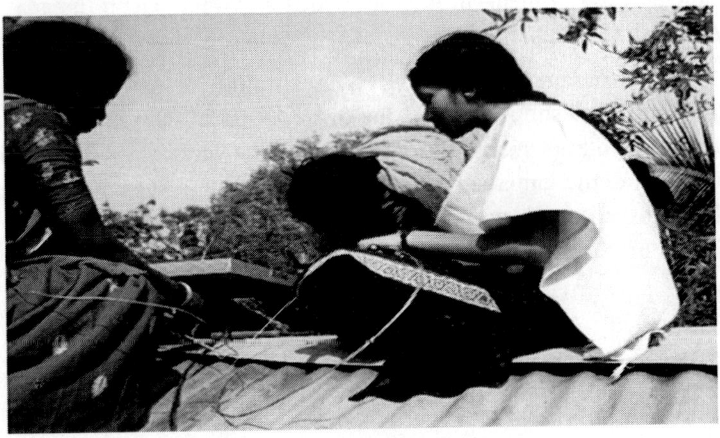

In the early hours of 27th morning the President of IFAD arrived and was happy to inaugurate the solar electrification project and to see the developmental activities taking place in the livelihood programmes financed by the organization itself.

The IFAD president was keen to know how the programmes on energy are related to livelihoods. An explanation providing an answer to the query was provided in which it was emphasized that 46 % of the households in rural India are not electrified which explains its reason for backwardness in the country. Once energy is provided to the poor, there is a possibility for the social, economic, human and financial development in the individuals, which again means a livelihood development.

This programme is worked with a future to set up a Rural Electronic Workshop which is due to be set up in the village with a capacity to attract many entrepreneurs in the technology who will manufacture the lanterns or the solar home lighting systems in other villages based on the projects received from the village or from other villages or states of the country.

4

Impact of Solar Electrification Projects

4. Introduction

Despite all the efforts made by the Government in developing the skill of the women on solar Electrification, WBSEA has been very back ward in marketing its products. The products so manufactured are sold within the NIRD campus and not much business was made possible giving in financial returns. Many organizations and individual entrepreneurs were coming up and requesting for costing details of the systems based on their need, day and night the project details were perused and the costing details were provided to the enquirers but to the utter dismay of WBSEA none responded to the replies given and none came up to give a project to WBSEA. The reason being the expensive nature of the solar systems, despite being environment friendly and a simple technology to handle. Against the understandings, this chapter gives us insights into how WBSEA identified and developed collaborative linkages with other organizations and learnt from their experiences in dealing with the business in general and solar energy in particular through activities including organization and participation of (in) seminars and workshops aimed at sensitizing policymakers about the effectiveness of the barefoot approach.

4.1 The TITAN Experience for Marketing WBSEA Products

In one of the one day workshops the Titan Company displayed its products as they were also the participants of the programme.

Titan is one of the companies involved in the manufacture of the solar panels. This company has also started producing solar lanterns at a lower cost than WBSEA. Its costing details confused WBSEA and they were quoted at a very low price. Upon analysis it was found that TITAN Company has provisioned an amorphous panel instead of silicon panel to power the solar lantern. The amorphous panel costs Rs .65 per watt and its life is proven to be not more than 9 years, additionally it is also expected that its efficiency decreases with usage and as promised the lantern's efficiency is not more than an hour as a radio is also fitted with the lantern. On the other hand the mono crystalline silicon cells cost Rs.210 per Watt and its life is tested and proven to be for 20 years. But in the market, in a bid to market the products the manufacturers are selling the lanterns by combining the amorphous panels and thereby lowering the sale price. Such interventions made the survival of WBSEA difficult as the products were not getting marketed. This gave lessons to WBSEA to be aware of the quality of the products and lobby for the quality of the product with the customers.

4.2. TERI Experience for Marketing WBSEA Products

The Deputy Director General in one of his visits to New Delhi happened to discuss on the solar technology with the professors of Tata Energy Research Institute (TERI). The DDG discussed the marketing interventions of TERI with WBSEA. Experiences of TERI in marketing their solar lanterns excited the WBSEA to replicate its model. Accordingly the deputy director general arranged for a meet with one of the, Associate Professor Mr. Ibrahim at New Delhi, this meeting enabled WBSEA to Compare the products prepared by WBSEA and TERI, it was seen that there are no virtual differences between the products produced by TERI and WBSEA in the context of costing and material input. But TERI's approach in terms of marketing has been through the selection of dealers in prominent villages and towns, who sold out the lanterns and provided servicing facilities to the customers. TERI provides training to the dealers so as to handle the problems associated with the servicing of the lanterns. TERI also took the responsibility to advertise their product in the villages and build awareness in the people on utilization of solar energy systems through play lets and songs.

Initially TERI gave the lanterns in bulk to the dealers. Nearly 5-6000 lanterns were sold out to the dealers in a month, but to the surprise of TERI the lanterns were meddled and returned as there was a guarantee period for 2 years, this put TERI on loss. This experience discouraged WBSEA from replicating such marketing technologies for itself. Gradually WBSEA started gaining recognition and its sales too started picking up slowly.

4.3 Linking up with International Development Center for a Marketing Avenues

To further explore the possibilities of marketing WBSEA products, the members of WBSEA attended a two day deliberation on energy alternatives for rural development held at New Delhi from 27-28th April 2006. The seminar was inaugurated by the Chief Guest Shri U N Panjiar, Additional Secretary, Ministry of Power, Govt. of India, and the Guest of Honor was Shri Anil Kumar Lakhina, C M D Rural Electrification Corporation Limited, New Delhi and the keynote address was provided by Shri R K Jain, Director (Technical), NTPC, New Delhi.

The participants presented innovative papers from renowned research organizations like Dept. of Elect Engg, IIT Delhi, Tata Energy Research Institute (TERI), IIT Roorkee, Uttaranchal Power Corporation Ltd. (UPCL) and Dehradun. And big corporates like Asian Electronics Ltd New Delhi, Indo German Energy Programme (IGEN) etc, participated and discussed on renewable energy technologies related to sun, wind, bio and hydropower based systems and on the research work and technologies developed by scientists. Here WBSEA had an opportunity to present a paper in the first session scheduled on 'Energizing Rural Sector through Grid ', which was chaired, by Padma Bhushan Shri G L Tondon, former CMD, Coal India. The executive member WBSEA Dr G Valentina, represented the National Institute of Rural Development, Hyderabad, and made a presentation on the unique approach of WBSEA on "Sustainable Energy Development through Peoples Participation - A

Case of Women Barefoot Solar Engineers of Andhra Pradesh ", here all the activities of WBSEA through the barefoot approach at Rural Technology park, NIRD was elaborated to the audience in the workshop. The participants of the workshop appreciated the presentation and the work done by WBSEA and sought details on the possibilities of association of WBSEA with other institutions for transfer of solar energy technologies on a barefoot approach in the rural areas. The chairman Padma Bhushan Shri G L Tondon, former CMD, Coal India too appreciated the presentation and sought the details on WBSEA approach to incorporate it as a policy proposal in the 10th plan and also the ensuing energy programmes on Rajiv Gandhi Grameen Vidhuthikaran Yojana in various parts of the country.

WBSEAs barefoot approach was selected as the best presentation for sustainable approach for rural energy development in the country. WBSEA thus bagged the first prize for its presentation. But nothing really happened in terms of linkages for marketing WBSEA products.

4. 4 Monitoring and Evaluation of Technology Transfer in Thamingula and Poosalapalam Villages

After the transfer of technology in the village a small monitoring and evaluation study was conducted in the Technology Transferred villages of Thamingula and Poosalapalem and it was observed that the people in the village seemed happy and gave an impression that the village is prospering. There were no maintenance problems even after two years of its installation in the village and all the systems were in good condition. All the houses were having music systems and were working through the solar power systems. Economic activity was also in full swing. At the time when NIRD team visited the village the villagers were cropping the turmeric crop (Paspu) and the village cropped nearly 300 - 500 kg of turmeric which was sold in the market @ Rs. 45/ kg , in addition they cropped 30-38 kg of French beans (Chi kudu) which was marketed @ Rs. 30 per kg). While 20 - 25 bags of Paddy was cropped on an average by each house hold which was sold @ 60-70 kg.

All the houses were illuminated with the solar lights in the night and also during the day time. Children had the opportunity to stay at home and study till late in the evening and complete all their home works which otherwise was not possible before solar electrification of the house. Even women could work cheerfully during the night times and prepare for the next day. They also felt secured and free in their mind from the possible attacks from poisonous insects and reptiles.

Most of the household possessed livestock and used them for irrigation purposes. In addition chicks were reared for sale and consumption. It was encouraging to learn that the house holds diverted the money used for the purchase of kerosene to address the educational needs of the children. But it was disappointing to learn that some of the families (20 %) used the money for drinking. One of the households has a television in the village and all the villagers assemble near the house to watch the news in the TV in the evenings, while the youth watch the cricket matches when aired.

4.5 Problems in the Project

After the completion of one year of the installation of the solar energy systems in the village, it is a happy note to observe that the systems were working in a good condition and every household in both the villages illuminate their houses with solar energy systems. The scene of illuminated households in the night made WBSEA very happy and satisfied with the work done by them. But as the project took off, various problems started surfacing and in the process these problems frustrated the continuation of the programme sometimes, while at certain times it was challenging and could be handled very effectively. By and large the problems are inherent in any system and the success of the programme lies in addressing them effectively.

4.6 Problems in the Implementation of the Project

There are many problems which can be documented in the project but a few of them are highlighted to understand its nature. Firstly the finances are not released by the concerned authorities and WBSEA had to finance for the project on its own, despite which the officials take years to release the finances which means a loss in terms of interest that WBSEA can accrue from the bank on the money that has been utilized which otherwise would have remained in their account. This loss is counted as huge money, as the funds were not released immediately and monitoring and evaluation of the systems also was not possible for WBSEA, at least until one year. This not only delays the project process but interferes with the subsequent evaluation and the timely incorporation of the changes also cannot be attended to as and when required and there is a danger of the project going off the track and achievement of the goals may be questioned.

Secondly, after setting the norms of payment and management of the systems through community participation, the officials in their visits have changed all the norms set by WBSEA. Especially in the payment of user charges. These payments were agreed upon by the community for sustaining the systems in the future which may be in terms of the payments that arise, to change the batteries, any essential parts of the system and payment of service charges to the trainees for effecting the servicing in times of need. WBSEA, through community participation agreed to collect Rs.100 as user charges for which the VEEC is the custodian of the funds. But the officials in their bid to express their so called empathy, who went to inspect the functioning of the systems and started making queries as to whether they can pay the amount decided by the community as user charges, recognizing the mood of the official, the community immediately retorted that it is not at all feasible for them to pay Rs.100 but instead will pay Rs.50 per month as user charges .Without a scientific understanding of the calculations made, they took decisions and proclaimed that the community may pay how much ever they can pay. This issue came up for discussion on two occasions when the state government officials visited the villages. WBSEA witnessed the discussions through recorded documents when WBSEA visited the village after an year.

There was confusion created in the village and the villagers reported that the state government official has asked them to pay only a sum of Rs.100 while the project director had asked them to pay how much ever they can. WBSEA was astonished to find that the whole project was jinxed up and became very apprehensive that if the user charges do not flow into the account, sustainability of the project is doomed to be a question as a result of such a confusion, when the VEEC and the trainees went to collect the user charges, the people turned hostile towards them. The villagers even went to the extent of abusing the trainees personally. It is almost a year and not single paisa was collected in the account from user charges.

WBSEA had to spend two days extra to make the people understand the amount of finances required to sustain the project. When all the accounts were analyzed before the households they understood the statistics and the need to make the payments. It is not that the households are poor or that they cannot pay the amount. The households can pay the amount but as the officials are showing their concern they aim at getting the payments forfeited and get the systems free of cost, since it is the first time that they are making any such payments for developmental activities which otherwise was achieved free of cost by them.

4.7 Problems in the Collection of user Charges

The VEEC is given the charge to collection the user charges from the households. But the households do not have any responsibility or reverence for the community. Whenever the VEEC requested the remittance of the user charges, some of the members of the village pulled off the systems and brought and threw the wires before the trainee and complained that these systems do not work, and they do not need such systems. Tolerating all their hysteric behavior the trainees reinstalled the system and found no fault in the system. In the VEEC meeting when WBSEA questioned the delay in the payments each one of them answered that they are waiting for the other to make the move first. Even the sarpanch who is a member of the VEEC has neither made any payments nor mobilized the households to make the payments. He has also declined from setting an example to the villagers that he has made the payments which can set blinkers on the mindset of the people to make payments when others start paying. With a great difficulty the households have agreed to pay Rs.70 per year as user charges .They have also agreed to pay the last years due also at the same rate but all these juggles brought down the remuneration of the trainees drastically. This will certainly demotivate the trainees to further provide any services in the future and there is always a danger that if the trainee refuses or does not show any interest in extending the services, the youth may mishandle the systems and damage them beyond the comprehension of VEEC or WBSEA to provide for either a replacement or some services on the system.

The solar system has relieved the households of Rs.100-150 per month towards kerosene expenditure, but the households are spending the money in boozing more but not in making payments for the systems.

4.8 Misuse of the Systems

Two to three instances of misuse of the systems are seen in the village wherein the households have let out the batteries for rent to run a music system or community lighting on occasions. The owner though he does not earn much in the game he only gets a bottle of drink for giving out his system. In such indulgences his battery will discharge faster and he will have to change the system earlier than the others. Knowing the misuse, the VEEC may not agree to change the battery but it has an occasion to blame the quality of battery installed by WBSEA. In another occasion a house hold which has bought a television has stealthily connected it to the system in the workshop which is enabling the whole village to watch the television for the whole night or during the whole daytime on Sundays and also to watch the cricket matches, this again has an impact on the batteries in the future.

In order to monitor all the issues discussed, an intervention of village level officials or an NGO is essential lest the responsibility rests on WBSEA to collect the user charges and monitor the systems against malpractices, which is located at a distance of 700-800 km.

4.9 Lack of Consensus Amongst the Trainees

The trainees in the village also lack a consensus in their decisions and work. One of the trainees called Papayamma took all the responsibility to provide servicing to the systems that had trouble, while the second trainee called Pravalika neglected the work since she is not paid the service charges so she will not be able to take up the job. She was also in a notion that Papayamma is trying to monopolize the whole activity and she is side tracked and no role to play. When WBSEA received the complaint, she had to be explained that her remuneration will depend on the services that she is going to provide. Then WBSEA made arrangement to provide a duplicate key to the other BSE so that she can carry out the work in the absence of the other, as this was a major grouse in the BSEs.

4.10 Idling of Workshop

The workshop has been installed incurring an expenditure of Rs 2 lakhs for the village to carry out the servicing work and also income generation activities in the village. The workshop is also provisioned with a community Television but as the finances are yet to be released from the government of Andhra Pradesh. The VEEC has collected more than 2 lakh rupees which can be used to purchase the solar material and the youth of the village can market the lanterns as there is a demand in the village for 10w 12 volt solar lantern. But such interventions are not sparking off from the village due to lack of initiative from the people and also the officials.

4.11 Intervention of the State Government Through RGVY

In its premier attempts to electrify all the inaccessible villages in the country enunciated through a policy at the centre the project villages Thamingula and Poosalapalem also were considered for electrification. It was disheartening since all the efforts made to sustain the villages on energy security were sapped off. However the current so provided is not quality current and the households alternate the usage of power whenever the grid power fails. But how about the sustenance of the systems and all the money (in crores of rupees) spent to electrify these villages? Does the government work this way duplicating its efforts and financial allocations?

Summary and Conclusion

The book provides action research interventions primarily on Promotion of Solar electrification projects and application of solar energy technologies for rural development which constitutes one of the most important interventions for rural development since energy access plays a very dominant role in development of the community. People depend on energy for myriad activities from lighting their households to that of cashing in on economic activities which has a potential in generating quick results and according easy comfort to mankind. But access to energy has remained a scarce source to a total non availability especially in the tribal areas due to the high installation costs and non extension of concerted maintenance services. Previous experiences have established that all efforts in according the services to the rural poor amounted to futility in the efforts of the government despite pumping in cores of rupees in facilitating the Solar based infrastructural facilities.

These disappointments were overcome by strategizing a project incorporating all events on the activity involved from selection and networking semi-literate poor rural women from non-electrified villages/hamlets as barefoot solar engineers (BSE) for undergoing training and capacity building on supply and demand aspects of the electrification model. The technology application can be seen from two angles; firstly in gaining energy security for the poor and management of the systems by the community. Secondly Solar Energy technology can also provide further livelihood support to the rural poor by according projects on a turnkey basis in the rural entrepreneurship development, since Solar technology is a promising income generating venture for the rural poor and capacity building on fabrication of the technology components like charge controllers and inverters, assembly of solar lanterns, fixed power packs in a Rural Electronic Workshop and marketing of the finished products and its related services go a long way in pronouncing this avenue of the programme. Synergizing the technology with women self help groups is an overarching feature of the project. This link up is advantageous for the community which anchors on the saying - teach a women and she will teach the whole society but teach a man he will keep it with himself. Thus extension of the technology through the women SHG members has been the prime goal of the project aiming at sustainability.

Thus the project evolved and is placed on the following four pillars;

1. Peoples participation
2. Capacity Building
3. Women empowerment and
4. Sustainability of the technology

In conclusion the emerging model on the action research has proved that semi-literate rural woman have from remote village has the capacity and competence to solar electrify her own village and a solar engineer is not required for the purpose. Secondly, those communities in non-electrified villages must be involved in planning and implementation process and decision-making which must be collective and transparent to make it sustainable. Thirdly, this enables the rural women not only to increase their income by providing the services but also their self-respect and self-esteem are enhanced by being identified as engineers. The first four Barefoot Engineers increased their income from Rs 500/month to Rs 3,000/ month bringing more money home than their husbands. Fourthly, it is possible to transfer the management, control and ownership of the most sophisticated technologies to the remotest of rural areas. Fifthly, this approach is cost effective and replicable. It involves an equal partnership between governments, NGOs and the communities where the financial, technical and management roles are clearly defined. Sixthly, by providing solar lighting it has led to a substantial decrease in consumption of fossil fuels like kerosene and diesel and made them self-sufficient in power and seventhly, by using solar power for lighting it was possible to demonstrate how solar energy could also be used for heating, drying, cooking and village level income generating activities.

However it is observed that the solar energy technologies are not able to penetrate in the desired direction and intentions of the policy makers for the following reasons; firstly the technologies are expensive therefore affordability of the poor to such technologies for energy access / security is a moot point. Therefore there is always a dependence of the community on the government agencies to release the technologies and implement them free of cost. Therefore mobilization of the poor for a contribution needs a careful game play in the process. Subsequent entrepreneurship development depends on reception of the projects on a turnkey basis and striking a deal to gain access to the projects is not in the ambit of the poor since it involves the intervention of the educated personnel. Therefore handing over the technology totally into the hands of the poor for entrepreneurship development may not give the desired results. Thus an appropriate involvement of an enterprising NGO, Panchayat or a private agency residing locally is the need of the hour to sustain the solar electrification projects in the future.

Annexures

Annexure-1

Sequence	Name/ Age	Husbands Name/ Family Size	Address	Present Occupation	Languages Known	Past Experience	Educational status
First	Smut Chena 45years	Sri.Sailu 4children	H.NO. 15,Vijaya Sagar Colony, Rajendera Nagar Hyderabad 500030	Barefoot solar Engineer	Telugu Understand Hindi	Stone cutter, House maid	Ittite-rate
Second	Smt Zaheda 30 years Vice president	D/O Sayed Ghulam Ali	Qrt.No.25 NIRD Campus, Rajendra nager, Hyderabad 500030	Barefoot solar Engineer	Telugu, and Hindi	Worked as Dyer in Cloth Factory	4th Standard
Third	Smt. Kala vathi 32 years General Secretary	S.M utthiah 3 Children	Qrt.No. A-25, Rajendra nagar Hyderabad 500030	Barefoot solar Enginer	Speaks Telugu and Hindi	Worked as Maid servant worked in Khadi Stores	5th Standard

Fifth	Smt. Yellam-ma 25 years Treasurer	Shri. Ramesh 4 children	H.NO.3 Vijay sagar Colony, Rajendra nagar Hyderabab 500030	Barefoot solar Engineer	Telugu understands Hindi	Stone Cutting	Illiterate

Executive Members							
Sixth	Smt. M.Kaseamma 20 years	W/O M.Yellappa	H.no.-17 Vijay Sager Colony, Rajendranager	Self Employed	Speaks Telugu	Stone cutter	Illiterate
Seventh	Smt.Assamma	W/O Assappa	H.no.-17 Vijay Sager Colony, Rajendranager	Self Employed	Speaks Telugu	Waste picker	
Seventh	Dr.G.Valentina	-	H.no.9-2-509/B, Langer House, Hyderabad	Assistant Professor NIRD	English, Hindi, Telugu,	Faculty	MA, MPhil, PhD.
	Dr.N.Madhuri			Assistant Professor NIRD	English, Hindi, Telugu	Faculty	M.A,PhD

Anexure-2

The village energy and environment committee (VEEC) of poosalapalem village comprises of the following members.

1. Sgina Peddabbai (ward member)-35 years.
2. Khather Somullaiah-40 years
3. Sagina Bhimaraju-32 years
4. Pangi chitti babu-35years
5. Lotha Atchi Raju-30 years.
6. Pangi balamma W/O Nageshwar rao-28 yrs.
7. Sagina Satyavathi W/o Babu Rao -28 yrs

8. Dukkeri Laxmmma-28 yrs. W/o Kondanna
9. Sagina Narayanaamma-35 yrs- W/o Laxmiah
10. Sagina Laxmi w/o Venkateshwarlu -25 yrs
11. Korrapalu W/o Raja Rao- 38 years
12. Sagina Kondamma-28 yrs- W/o Kondiah
13. Sagina Laxmi W/o Matsya lingam- 25 yrs.
14. Pangi Pandamma W/O Rambabu-25 years

Annexure-3

Executive Members							
Sixth	Smt. M.Kaseamma 20 years	W/O M.Yellappa	H.no.-17 Vijay Sager Colony, Rajendranager	Self Employed	Speaks Telugu	Stone cutter	Illiterate
Seventh	Smt.Assamma	W/O Assappa	H.no.-17 Vijay Sager Colony, Rajendranager	Self Employed	Speaks Telugu	Waste picker	
Seventh	Dr.G.Valentia	-	H.no.9-2-509/B, Langer House, Hyderabad	Assistant Professor NIRD	English, Hindi, Telugu,	Faculy	MA, MPhil, PhD.
Eighth	Dr.Madhuri	-		Assistant Professor NIRD	English, Hindi, Telugu	Faculy	M.A, PhD.
Ninth	Bagwathnanadanji	-	Social Work and Research Centre, Tilonia Rajasthan.	Coordinator-Solar Energy	Englis, Hindi, Rajasthani	Faculy	Intermediate
Tenth	Satayanarayan	-	Social Work and Research Centre, Tilonia Rajasthan.	Master Trainer-Solar	English, Hindi, Rajasthani	Faculy	Intermediate
Elevenh	Mrs.S. Papayamma	Krishna	Thamingula, chintapalli (M), Paderu (D) Vishakapatnam	Tribal Trainee BSE	Telugu	Traine	4th class
Twelve	Pravilika	Ramesh.	Thamingula, chintapalli (M), Paderu (D) Vishakapatnam	Tribal Trainee BSE	Telugu	Traine	Illiterate
Thirten	Pangi Kumari–	ChittiBabu	Poosalapalem, chintapalli (M), Paderu (D) Vishakapatnam	Tribal Trainee BSE	Telugu	Traine	2nd class.

The following are the names of the members of VEEC in Thamingula Village:

1. Pottur Raja Babu- -32 years
2. Pottur Bangaru Babu - 35 years - Vice President
3. Bangaraiah - 42 years - Sarpanch
4. Krishna - 25 years
5. Ramu - 26 years
6. Kondamma - Wife of Simhachalam - 25 years
7. Jattalingamma - W/o Linganna - 30 years
8. Pottur Lakshmamma - W/o Kondababu - 40 years
9. Pottur Chilakamma - 35 years - W/o Ramannadora
10. Pangi Thulasi - 25 years - W/o Damanaiah
11. Salamithi Rajulamma - W/o Rama Rao - 35 years
12. Sundru Chinna Bullemma - W/o Bangaraiah - 40 years
13. Salamithi Bangaramma - W/o Sanyasaiah - 55 years
14. Sundra Chinnatalli - Wife of Subba Rao - 27 years
15. Devudamma - 38 years

Annexure 5

A. financial allocation received from ITDA paderu by WBSEA:

Sno	Cheque received on	Amount	Utilization
1.	22.12.05	5.0 Lakhs	Spent for procurement of components for solar home Lighting
2.	16.2.06	2.5 Lakhs	-do-
3.	23.3.06	10 Lakhs	do-
	Total	17.50 Lakhs	

Amount received for installation of Home lighting is Rs. 17.50 Lakhs and sanction towards the same is Rs. 17,84,329 lakhs. Thus Rs. 17,84,329- Rs. 17.50 = Rs. 34,329 is due.

B. Following amount has to be released by ITDA -Paderu against completion of Phase -I of the project:

S no	Purpose	Amount to be released	Remarks	Met the finances from
1.	BSE Training	Rs.2, 40,765	Completed	WBSEA Account
2.	Installation of home lighting system	Rs. 34,329	Completed installation	WBSEA Account
3.	Rural Electronic Workshop	Rs.2, 09,277	Completed installation	WBSEA Account
4.	VEEC Training	Rs.40, 000	Completed	WBSEA Account
	Total	Rs.6, 24,371		

Final Picture of the allocation and Release of the Finances towards Phase-1

Sno	Total amount sanctioned in phase –I of the project	Total amount released as on date for completion of phase	Total amount to be released in phase-I of the project is	Remarks
1.	Rs.23,34,371	Rs.17.5 Lakhs	Rs.5, 84,371	This amount is inclusive of Rs. 60,000 to be released in the month of June 2006 towards monitoring and evaluation of the project.